T0140275

Studies in Computational Intelligence

Volume 848

Series Editor

Janusz Kacprzyk, Polish Academy of Sciences, Warsaw, Poland

The series "Studies in Computational Intelligence" (SCI) publishes new developments and advances in the various areas of computational intelligence—quickly and with a high quality. The intent is to cover the theory, applications, and design methods of computational intelligence, as embedded in the fields of engineering, computer science, physics and life sciences, as well as the methodologies behind them. The series contains monographs, lecture notes and edited volumes in computational intelligence spanning the areas of neural networks, connectionist systems, genetic algorithms, evolutionary computation, artificial intelligence, cellular automata, self-organizing systems, soft computing, fuzzy systems, and hybrid intelligent systems. Of particular value to both the contributors and the readership are the short publication timeframe and the world-wide distribution, which enable both wide and rapid dissemination of research output.

The books of this series are submitted to indexing to Web of Science, EI-Compendex, DBLP, SCOPUS, Google Scholar and Springerlink.

More information about this series at http://www.springer.com/series/7092

Roger Lee
Editor

Computational Science/Intelligence and Applied Informatics

 Springer

Editor
Roger Lee
Software Engineering and Information
Technology Institute
Central Michigan University
Mount Pleasant, MI, USA

ISSN 1860-949X   ISSN 1860-9503   (electronic)
Studies in Computational Intelligence
ISBN 978-3-030-25227-4   ISBN 978-3-030-25225-0   (eBook)
https://doi.org/10.1007/978-3-030-25225-0

This Springer imprint is published by the registered company Springer Nature Switzerland AG
The registered company address is: Gewerbestrasse 11, 6330 Cham, Switzerland

Foreword

The purpose of the 6th ACIS International Conference on Computational Science/Intelligence and Applied Informatics (CSII 2019) which was held on May 29–31 in Honolulu, Hawaii was to together researchers, scientists, engineers, industry practitioners, and students to discuss, encourage, and exchange new ideas, research results, and experiences on all aspects of A Computational Science/Intelligence and Applied Informatics and to discuss the practical challenges encountered along the way and the solutions adopted to solve them. The conference organizers have selected the best 13 papers from those papers accepted for presentation at the conference in order to publish them in this volume. The papers were chosen based on review scores submitted by members of the program committee and underwent further rigorous rounds of review.

In chapter "The Analysis on Commercial and Open Source Software Speech Recognition Technology", Jong-Bae Kim and Hye-Jeong Kweon compared and analyzed features and functions of typical speech recognition software in the commercial and open-source software fields. It is expected that the comparison and analysis on the features and functions of commercial and open-source software in the speech recognition software field carried out in this study could suggest a standard for selecting speech recognition technologies to help use proper API in context.

In chapter "Implementation of Electronic Braille Document for Improved Web Accessibility", Ho-Sung Park, Yeong-Hwi Lee, Sam-Hyun Chun and Jong-Bae Kim discuss the problems associated with the access to electronic documents that are provided by an information system through the Web or email are examined from the perspective of information access of visually impaired people. They propose a method to convert data of an electronic document into braille in the information system server for enabling visually impaired people to access documents more accurately.

In chapter "A Reliable Method to Keep Up-to-Date Rights Management Information of Public Domain Images Based on Deep Learning", Youngmo Kim, Byeongchan Park and Seok-Yoon Kim propose a reliable method of integrating the RMI representation system and updating the RMI with the up-to-date information based on the most reliable data among information collected from each site through a comparative search technique for public domain images based on deep learning.

In chapter "Applying GA as Autonomous Landing Methodology to a Computer-Simulated UAV", Changhee Han proposes a genetic algorithm method to achieve autonomy of unmanned aerial vehicles and will check the possibility of self-regulated autonomous unmanned aerial vehicle by applying the genetic algorithm.

In chapter "Vision-Based Virtual Joystick Interface", Suwon Lee and Yong-Ho Seo propose a virtual joystick system, which is a type of virtual input device. Their system detects a handheld stick and computes the direction in which the user's hand moves relative to a user-defined center. The proposed system's accuracy is competitive and has real-time speed in the laptop environment.

In chapter "A Study on Improvement of Sound Quality of Flat Display Speaker by Improving Acoustic Radiation Characteristics", Sungtae Lee, Kwanho Park and Hyungwoo Park analyze the acoustic characteristics of a flat speaker to realize such stereophonic sound and improve sound quality for organic light-emitting diode (OLED) panel televisions.

In chapter "Edge Detection in Roof Images Using Transfer Learning in CNN", Aneeqa Ahmed, Yung-Cheol Byun and Sang Yong Byun employ CNN method to detect edges of roof images. Incorporating CNN into edge detection problem makes the whole system simple, fast, and reliable. Moreover, with no more extra training and without additional feature extraction, CNN can process input images of any size.

In chapter "Improvement of Incremental Hierarchical Clustering Algorithm by Re-insertion", Kakeru Narita, Teruhisa Hochin, Yoshihiro Hayashi and Hiroki Nomiya attempt to improve the incremental clustering method. By examining the cluster multimodality which is the property of a cluster having several modes, they can select some points of a different distribution inferred from a dendrogram and transfer the points in the cluster to a different cluster.

In chapter "A New Probabilistic Tree Expression for Probabilistic Model Building Genetic Programming", Daichi Kumoyama, Yoshiko Hanada and Keiko Ono propose a new expression of probabilistic tree for probabilistic model building GPs (PMBGP). Tree-structured PMBGPs estimate the probability of appearance of symbols at each node of the tree from past search information and decide the symbol based on the probability at each node in generating a solution. Through numerical experiments, they show the effectiveness of the proposed probabilistic tree by incorporating it to a local search-based crossover in symbolic regression problems.

In chapter "Infrastructure in Assessing Disaster-Relief Agents in the RoboCupRescue Simulation", Shunki Takami, Masaki Onishi, Itsuki Noda, Kazunori Iwata, Nobuhiro Ito, Takeshi Uchitane and Yohsuke Murase propose a combination of an agent development framework and experiment management software in this study as infrastructures in assessing disaster-relief agents in the RoboCupRescue Simulation. In the evaluation, a combinatorial experiment as a case study confirms the effectiveness of the environment and shows that the environment can contribute to future disaster response research that utilizes a multi-agent simulation.

In chapter "OOCQM: Object Oriented Code Quality Meter", Asma Shaheen, Usman Qamar, Aiman Nazir, Raheela Bibi, Munazza Ansar and Iqra Zafar propose a framework named Object Oriented Code Quality Meter (OOCQM) for measuring source code quality of object-oriented code using low-level code metrics and high-level quality factors.

In chapter "A Fault-Tolerant and Flexible Privacy-Preserving Multisubset Data Aggregation in Smart Grid", Hung-Yu Chien and Chunhua Su propose a new PPMA scheme that facilitates flexible SM deployment, independent SM status reporting without strict synchronization, and fault tolerance to any SM failure as long as at least two well-function SMs.

In chapter "Secure and Efficient MQTT Group Communication Design", Hung-Yu Chien, Xi-An Kou, Mao-Lun Chiang and Chunhua Su design a secure MQTT group communication framework in which each MQTT application would periodically update the group key and the data communication can be efficiently and securely encrypted by the group keys. Both the prototype system and the analysis show that our design can improve the performance of security, computation, and communication.

It is our sincere hope that this volume provides stimulation and inspiration and that it will be used as a foundation for works to come.

May 2019

Hitoshi Iima
Kyoto Institute of Technology
Kyoto, Japan

Contents

Contributors

Aneeqa Ahmed Department of Computer Engineering, Jeju National University, Jeju-si, Republic of Korea

Munazza Ansar Department of Computer Engineering, NUST, College of E&ME, Rawalpindi, Pakistan

Raheela Bibi Department of Computer Engineering, NUST, College of E&ME, Rawalpindi, Pakistan

Yung-Cheol Byun Department of Computer Engineering, Jeju National University, Jeju-si, Republic of Korea

Sang Yong Byun Department of Computer Engineering, Jeju National University, Jeju-si, Republic of Korea

Mao-Lun Chiang Department of Information and Communication Engineering, ChaoYang University of Technology, Taichung City, Taiwan

Hung-Yu Chien Department of Information Management, National Chi-Nan University, Puli, Nantou, Taiwan, R.O.C.

Sam-Hyun Chun Department of Law, Soongsil University, Seoul, Korea

Changhee Han Department of Computer Science, Korea Military Academy, Seoul, South Korea

Yoshiko Hanada Faculty of Engineering Science, Kansai University, Osaka, Japan

Yoshihiro Hayashi Research and Development Department, NITTO SEIKO CO., LTD, Ayabe, Japan

Teruhisa Hochin Information and Human Sciences, Kyoto Institute of Technology, Kyoto, Japan

Nobuhiro Ito Aichi Institute of Technology, Toyota, Japan

Kazunori Iwata Aichi University, Nagoya, Aichi, Japan

Jong-Bae Kim Startup Support Foundation, Soongsil University, Seoul, Korea

Seok-Yoon Kim Department of Computer Science and Engineering, Soongsil University, Seoul, Republic of Korea

Youngmo Kim Department of Computer Science and Engineering, Soongsil University, Seoul, Republic of Korea

Xi-An Kou Department of Information and Communication Engineering, ChaoYang University of Technology, Taichung City, Taiwan

Daichi Kumoyama Graduate School of Science and Engineering, Kansai University, Osaka, Japan

Hye-Jeong Kweon Department IT Policy and Management, Soongsil University, Seoul, Korea

Sungtae Lee LG Display, Paju-si, Gyeonggi-do, Republic of Korea

Suwon Lee Department of Computer Science and the Research Institute of Natural Science, Gyeongsang National University, Jinju-si, Gyeongsangnam-so, South Korea

Yeong-Hwi Lee ATSoft Co., Ltd., Seoul, Korea

Yohsuke Murase RIKEN Center for Computational Science, Kobe, Hyogo, Japan

Kakeru Narita Graduate School of Information Science, Kyoto Institute of Technology, Kyoto, Japan

Aiman Nazir Department of Computer Engineering, NUST, College of E&ME, Rawalpindi, Pakistan

Itsuki Noda National Institute of Advanced Industrial Science and Technology (AIST), Koto-ku, Japan

Hiroki Nomiya Information and Human Sciences, Kyoto Institute of Technology, Kyoto, Japan

Masaki Onishi National Institute of Advanced Industrial Science and Technology (AIST), Koto-ku, Japan

Keiko Ono Department of Electronics and Informatics, Ryukoku University, Shiga, Japan

Byeongchan Park Department of Computer Science and Engineering, Soongsil University, Seoul, Republic of Korea

Ho-Sung Park Department IT Policy and Management, Soongsil University, Seoul, Korea

Hyungwoo Park School of Information Technology, Soongsil University, Seoul, Republic of Korea

Kwanho Park LG Display, Paju-si, Gyeonggi-do, Republic of Korea

Usman Qamar Department of Computer Engineering, NUST, College of E&ME, Rawalpindi, Pakistan

Yong-Ho Seo Department of Intelligent Robot Engineering, Mokwon University, Seo-gu, Daejeon, South Korea

Asma Shaheen Department of Computer Engineering, NUST, College of E&ME, Mirpur AJK, Pakistan

Chunhua Su Division of Computer Science, The University of Aizu, Aizuwakamatsu, Japan

Shunki Takami University of Tsukuba, Tsukuba, Ibaraki, Japan

Takeshi Uchitane Aichi Institute of Technology, Toyota, Japan

Iqra Zafar Department of Computer Engineering, NUST, College of E&ME, Rawalpindi, Pakistan

The Analysis on Commercial and Open Source Software Speech Recognition Technology

Jong-Bae Kim and Hye-Jeong Kweon

Abstract The speech recognition technology is one that converts acoustic signals obtained through a sound sensor such as a microphone into words or sentences. The speech recognition technology appeared in the 1950s and has been studied continuously, but it has not been popularized until the mid-2000s because of its low speech recognition rate. Recently, however, the speech recognition technology, which has been used limitedly only for specific applications so far, is being developed rapidly along with the proliferation of portable computing terminals represented by smart phones and the expansion of cloud infrastructure to support them. In particular, the speech interaction system combining this technology with artificial intelligence is attracting attention as a next generation interface, and has been used in various fields such as smart phones, smart TV and automobiles. Recently, Samsung Electronics has released 'Bixby' combining artificial intelligence and speech recognition, and a lot of companies such as Google and Naver have offered speech recognition technologies as open API. This paper compared and analyzed features and functions of typical speech recognition software in the commercial and open source software fields. It is expected that the comparison and analysis on the features and functions of commercial and open source software in the speech recognition software field carried out in this study could suggest a standard for selecting speech recognition technologies to help use proper API in context.

Keywords Speech recognition · Comparative analysis · Open source software · Open API · Artificial intelligence

J.-B. Kim (✉)
Startup Support Foundation, Soongsil University, Seoul, Korea
e-mail: kjb123@ssu.ac.kr

H.-J. Kweon
Department IT Policy and Management, Soongsil University, Seoul, Korea
e-mail: pskhj@naver.com

© Springer Nature Switzerland AG 2020
R. Lee (ed.), *Computational Science/Intelligence and Applied Informatics*,
Studies in Computational Intelligence 848,
https://doi.org/10.1007/978-3-030-25225-0_1

1   Introduction

The speech recognition technology is one that converts acoustic signals obtained through a sound sensor such as a microphone into words or sentences [1]. The speech interaction system combining this technology with AI (artificial intelligence) is attracting attention as a next generation interface, and has been used in various fields such as smart phones, smart TV and automobiles [2]. In particular, Google has recently released 'Google Assistant' which is an AI speech recognition assistant, and also offered the speech recognition functions applicable to automobiles and consumer electronics etc. as open API. Also in Korea, development for speech recognition has been done actively such as Naver Clova, which is a speech recognition AI launched by Naver on May 12, 2017, and Bixby released by Samsung Electronics [3].

This paper compares and analyzes the features and functions of typical speech recognition software in the commercial and open source software fields. It is expected that the comparison and analysis on the features and functions of commercial and open source software in the speech recognition software field could suggest a standard for selecting speech recognition technologies to help use proper API in context.

2   Related Studies

Recently, open source software has received attention in many fields, and as an "open source way" is applied in the form of "collaboration" to various creation processes and the spirit of "sharing" is expanded also in using the output, a new trend such as "open source hardware" and "open source content" has emerged also in the areas other than the software field. This trend is a result coming from the development and diversity of this society, and is because the open source philosophy and methodology provides an inspiration to many people who is trying to turn to the sharing economy paradigm breaking through the industrial age. The concept of such an open source software evolved mainly by developers in early day, and the basic ideology that reveals the software source code and guarantees freedom of distribution is based on the philosophy of "sharing" and "contribution" [4, 5]. Open source software and commercial one show a great difference in terms of the detailed features of development models and the differentiation of user communities and added values etc., and this difference sometimes leads also to causing a confrontation between the two sides. In other words, there is a great difference in the underlying philosophical implication (aiming to create value different from the traditional view within the product and service).

However, there is no great difference between proprietary software and open source one as an end product from a technological aspect unique to software application development. In addition, it is difficult to find a significant difference between the two also in the technologies (for example, design, architecture or specific implementation) applied to the product during development. It is the same for the speech recognition technology related to the speech interaction system that attracts attention as a next generation interface. Recently, open source technologies comparable to commercial software continue to cascade rapidly, and in particular, as various companies such as Google and Naver offer speech recognition technologies as open API, the interest is rising in the open source speech recognition software.

Speech recognition is to identify linguistic meaning from voice by an automatic method, and specifically, it is a process to enter speech waveform to identify a word or a series of words and extract its meaning, which is divided into five main stages such as speech analysis, phoneme recognition, word recognition, sentence interpretation and semantic extraction.

In a narrow sense, it often includes from speech analysis to word recognition. The ultimate goal of speech recognition is the realization of a full speech-to-text conversion to recognize speech by natural vocalization to accept as an executive command or enter as data into a document. It is to develop a speech understanding system to use syntax information (grammar), semantic information, task-related information and knowledge etc. to exactly extract the meaning of continuous speech or sentences as well as only to recognize words [6, 7].

Meanwhile, the open API is an API opened for everyone to use. It is a set like protocols and tools prepared to easily create any applications, and program developers could easily develop applications only with a few opened API without understanding detailed functions of the operating system. For the global speech recognition open API, there is Google's Cloud Speech API, and Kakao's Newtone and Naver's Clova Speech API are provided typically in Korea.

3 Comparison Between Commercial and Open Source Speech Recognition Software

3.1 Commercial Speech Recognition Software

(1) Dragon Mobile SDK

Dragon Mobile started the speech recognition service based on iPhone, Android, Windows Phone 7 and HTML in January, 2011. At that time, it had a technique to convert text to speech (30 languages), and also one to convert speech to text (8 languages: Korean, English, French, Spanish, Dutch, Norwegian, Swedish and Chinese) (Fig. 1).

Dragon Mobile SDK provides the voice encryption service and the spoken language analysis monitoring tool, which is possible to do 500 tests for 90 days, and after that, it could continue to use if converting into paid subscriptions.

The operation mode could be divided into speech recognition and speech synthesis as Table 1.

Fig. 1 Dragon Mobile SDK

Table 1 The operation mode of the dragon mobile SDK

Category	Procedure
Speech recognition	Send speech from the smart phone app → send after converting text in the Dragon Mobile server → output after receiving text into the smart phone app
Speech synthesis	Send text from the smart phone app → send after synthesizing speech in the Dragon Mobile server → output after receiving the speech file into the smart phone app

Tests to experience the development process actually utilizing it are possible, and it is as follows if describing the procedure (based on the iPhone development environment). First, join the Dragon Mobile home page, and then download the SDK. Then receive a key value via the registered email after requesting credentials on the Dragon Mobile home page. Subsequently, it is a structure of being able to do a test if opening the speech recognition sample project (Dragon Mobile Recognizer) in the SDK and replacing the following part of the DMRecoginzerViewController.m file with the received key value (Table 2).

To apply respective countries' languages, the following part should be modified (Table 3).

Similarly, the speech synthesis sample project (DragonMobileVocalizer.m) should also be compiled after modifying the key value and language according to the language-specific environment as below (Table 4).

To apply it to actual developer apps since then, it could be operated by adding packages (packages in the sample such as SpeechKit.framework), by modifying and adding the recordButton Action function of the sample file (DMRecognizerView-

Table 2 Changing the key value of the Dragon Mobile SDK

```
const          unsigned          char          SpeechKitApplicationKey[]          =
{ INSERT_YOUR_APPLICATION_KEY_HERE };
....................
[SpeechKit setupWithID: INSERT_YOUR_APPLICATION_ID_HERE
                 host: INSERT_YOUR_HOST_ADDRESS_HERE
                 port: INSERT_YOUR_HOST_PORT_HERE
             useSSL:NO];
             delegate:self];
```

Table 3 Changing languages of the Dragon Mobile SDK

```
switch (languageType.selectedSegmentIndex) {
    case 0:
        langType = @"en_US";
        break;
    case 1:
        langType = @"en_GB";
        break;
    case 2:
        langType = @"fr_FR";
        break;
    case 3:
        langType = @"de_DE";
        break;
    default:
        langType = @"en_US";
        break;
}
```

Table 4 Changing the DragonMobileVocalizer.m language of the Dragon Mobile SDK

```
vocalizer = [[SKVocalizer alloc] initWithLanguage:@"en_US" delegate:self];
```

Controller.m), the related other functions, the UI part according to the app environment and linking the event starting speech recognition with the recordButtonAction function in the case of speech recognition, and by linking the speech synthesis target text to the [vocalizer "text"] part of the speakOrStopAction function after the work such as modifying the speakOrStopAction function of the sample file (DMVocalizerViewController.m), the related other functions and UI in the case of speech synthesis.

As a result of applying the test in accordance with this procedure, the following pros and cons were drawn (Table 5).

Table 5 Pros and cons of the Dragon Mobile SDK

Pros	Cons
(1) Support iPhone, Android, BlackBerry, and Windows Phone7 (2) Support a lot of languages (8 languages speech recognition, 30 languages speech synthesis)	(1) Frequent server disconnections (the free test environment is a possible cause) (2) The recognition rate is not perfect when recognizing and synthesizing Korean speech (3) Difficult to convert speech to a different tone after speech synthesis (4) Fail to process English words if Korean and English words are mixed during speech synthesis

(2) *Google Speech Recognition API*

Google Speech Recognition API is rated as a powerful speech recognition module. With easy-to-use API, developers could convert audio to text by using Google Cloud Speech-to-Text in which a powerful neural network model is applied (Fig. 2).

This API recognizes more than 120 languages and dialects in response to the global users. It could implement voice command and control functions, do operations such as converting call center's audio to text, and use Google's machine learning technologies to process real-time streaming or pre-recorded audio.

In particular, the most advanced deep learning neural network algorithm is applied to audio in terms of machine learning utilization, so it could recognize speech with unparalleled accuracy, and as Google continues to improve the internal speech recognition technology used in Google products, it is expected that the accuracy of Cloud Speech-to-Text is also improved over time.

In addition, as known to recognize more than 120 languages and dialects, the Cloud Speech-to-Text could support global users because it recognizes more than 120 languages and dialects.

Additionally, inappropriate content could be filtered from every language's text result.

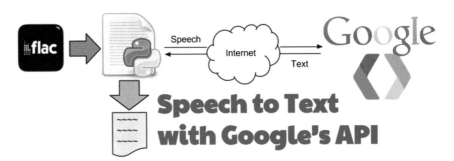

Fig. 2 Google speech recognition API

It also provides a function to automatically identify speech languages, and could verify which language is used in a specific speech if using the Cloud Speech-to-Text (currently, however, limited to 4 languages). This function could be used in voice search and commands, and return a result of converting short or long audio to text in real time. When streaming audio is recognized or a user is speaking, the Cloud Speech-to-Text could immediately return text to stream the text result, or the Cloud Speech-to-Text could return the text recognized from the audio stored as a file. In addition, any long audio as well as short one could be analyzed.

Meanwhile, Google Speech Recognition API could automatically convert a format suitable for the proper noun and context into text, and the Cloud Speech-to-Text is produced to properly recognize everyday conversation and is able to exactly convert proper nouns to text and specify the format of language appropriately. Additionally, Google supports proper nouns of more than 10 times compared to the total word count of Oxford English Dictionary.

Finally, Google Speech Recognition API provides the pre-built model collection suitable for use cases as Table 6 which provides advantages of being able to optimize in accordance with use cases (for example, voice command) because it is provided with many pre-built speech recognition models.

For example, the pre-built video text conversion model is suitable to create an index or caption of video or content which has several narrators, and uses the machine learning technology similar to YouTube captioning.

(3) *Siri*

Siri is an intelligent speech recognition personal assistant application operating on the Apple devices mounting iOS and macOS. It was unveiled with iPhone 4S on October 4, 2011, which supported only English, German and French at that time, but Japanese was added to iOS 5.1 update distributed in March 2012, and in iOS 6, more languages were added (Fig. 3).

If a user speaks, it is recorded to send to Apple's servers, and then the nuance technique is used to convert speech to text. And it is analyzed by artificial intelligence of SRI acquired by Apple to determine the operation. Accordingly, it tells answers or runs apps.

Table 6 Application models by use case of Google speech recognition API

Model	Description
command_and_search	Best fitted for short queries such as voice command or search
phone_cell	Best fitted for audio coming from telephone conversation, and the audio recorded at 8 kHz sampling rate usually is good
Video	Best fitted for audio coming from video or one with several narrators. Audio recorded at 16 kHz or more sampling rate is good. It is a premium model more expensive than the standard fee
Default	Best fitted for audio not specific audio models. For example, it includes long audio. Audio with Hi-Fi and recorded at 16 kHz or more sampling rate is good

Fig. 3 Apple's Siri

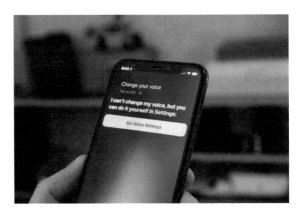

Siri, which provides functions such as dialing, playing music, text messaging and scheduling, supports iPhone, iPod Touch 5th generation or more, iPad 3rd generation or more, Apple Watch, Macintosh (devices mounting a version after macOS Sierra) and HomePod etc.

(4) *Yandex SpeechKit*

Yandex SpeechKit based on speech recognition and synthesis (text-to-speech conversion) technologies supports three languages such as Russian, English and Turkish. It has an advantage of being able to make natural speeches from more than a million individual phonemes with the audio and text processing and the intonation established in neural networks trained with a lot of real-life examples (Fig. 4).

The real-time synthesis technology of receiving the recorded audio in response to as soon as the service or app send a speech synthesis text is suitable to make software supporting audio streaming because its delay time is short. And, it has an advantage of being able to implement new functions quickly and easily with easy-to-use API without needing to provide HTTP and API for data exchange and to distribute and support its own infrastructure.

Yandex SpeechKit is suitable to implement functions such as automating appointment scheduling and making lectures and web seminars.

(5) *Microsoft Speech API*

Speech recognition   Speech synthesis   Voice activation

Fig. 4 Yandex SpeechKit

Fig. 5 Microsoft speech platform

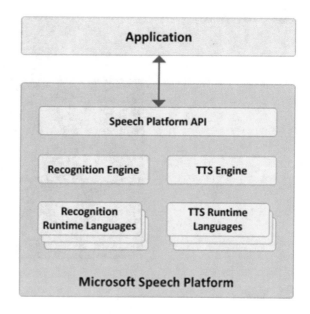

Microsoft Speech API has a multilingual text-to-speech conversion engine and multilingual speech recognition engine, which provides redistributable components that could make developers package the engines and runtime with application program codes to create single installable application (Fig. 5).

Microsoft Speech API provides a variety of SAPI API product family, and this product family could be applied to Microsoft narrator, Microsoft Office XP and Office 2003, Microsoft Excel 2002, Microsoft Excel 2003 and Microsoft Excel 2007 of Microsoft Windows XP Tablet PC Edition, Windows 2000 and Windows operating systems, and other Windows products.

3.2 Open Source Speech Recognition Software

(1) CMU Sphinx

CMU Sphinx is a general term that represents the speech recognition system group developed by Carnegie Mellon University. CMU Sphinx Group has led the tool and application development in the field of speech recognition related interaction systems and speech synthesis, financed by DARPA (Defense Advanced Research Projects Agency) (Fig. 6).

Sphinx is a speech recognition engine constructed with open source, and the PC version is divided into Sphinx-1~4, WinCE and Android, and the iPhone version is Pocketsphinx . The features of respective versions are as Table 7.

Fig. 6 CMU Sphinx

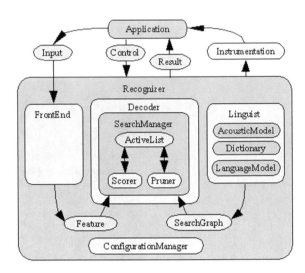

(2) *Kaldi*

Kaldi is a speech recognition toolkit offered free under the Apache License. The Kaldi speech recognition toolkit was released in 2011, which was implemented with C++ language based on OpenFST library, so its reusability and portability is superior, and well-documented. Researchers of various institutions with Johns Hopkins University in the United States and Brno University in the Czech Republic as the center are participated as programmers to actively carry out the version upgrade, and its source is opened through SourceForge. It supports most of the feature extraction and acoustic modeling algorithms widely used in the speech recognition field, and provides even experimental scripts using open speech database, so it is a great help also to the comparative study of recognition results. Recently, it supports even the deep neural network acoustic model, so many researchers are using this software. As it has the Apache license policy, it could be relatively free to use. It provides a lexical tree based tool developed by HTK to convert acoustic models (Fig. 7).

The Kaldi speech recognition toolkit provides several types of decoders. The SimpleDecoder is the most simple decoder, which finds the final recognition result based on the beam search algorithm, and is used for understanding the WFST based decoding algorithm or debugging the optimized decoder. The FasterDecoder quickly operates by using the pruning by likelihood and the double beam search limiting the number of states activated per frame, and is used for developing actual applications. In addition, the Online FasterDecoder for online speech recognition, BiglmFaster-Decoder that dynamically generates a network for a large language model to process it, and Lattice SimpleDecoder/LatticeFasterDecoder to use lattice to post-process the speech recognition result in multiple stages are provided as different separated programs.

Table 7 Features by version of CMU Sphinx

Version	Features
Sphinx	① Continuous speech, speaker independent recognition system using the HMM (hidden Markov acoustic model) and n-gram statistical language model ② Developer: Kai-Fu Lee ③ At that time of 1986, it featured the validity of continuous speech, speaker independent and large vocabulary recognition. The performance was replaced by the subsequent version
Sphinx 2	① It integrated functions such as endpointing, partial hypothesis generating and dynamic language model migration ② It was used in the interaction system and language learning system ③ It could be used in a computer based PBX system such as Asterisk ④ Currently, active development is not done except routine maintenance ⑤ Current real-time decoder development is carried out in the Pocket Sphinx project
Sphinx 3	① It adopted the continuous HMM representation ② It was used primarily in high-precision non-realtime recognition ③ The latest modeling technology such as LDA/MLLT, MLLR and VTLN was used with SphinxTrain to improve the recognition accuracy
Sphinx 4	① It was one that completely built the Sphinx engine again with the goal of providing the speech recognition research built by Java programming language as a more flexible framework ② Sun Microsystems supported the development of Sphinx 4, and provided the project with software engineering expertise (Individuals of MERL, MIT and CMU were included in the participants) ③ Current development goals ⇒ Develop a new (acoustic model) trainer ⇒ Speaker adaptation (ex. MLLR) ⇒ Improve configuration management ⇒ Generate graphics-based UI for graphic system design
PocketSphinx	① Sphinx version that could be used in an embedded system (ex. ARM processor based) ② Currently under active development ③ Integrate functions such as an efficient algorithm for fixed-point arithmetic and GMM computation

(3) *Julius*

Julius is a large vocabulary continuous speech recognition engine built with C language by Kyoto University in Japan, which provides the Windows/UNIX 20 k Japanese dictation kit. Julius is high-performance software with two large vocabularies continuous speech recognition (LVCSR) decoders for speech-related researchers and developers. It could carry out decoding for a 60 k words dictation task on most current PCs in near real time by using the word 3-gram and context-dependent HMM.

Major platforms are Linux and Unix workstations, and it also operates on Windows. Julius is distributed with source codes as an open source license (Fig. 8).

It is modularized independently from the model structure, and supports a variety of HMM types. It was developed as a part of free software toolkit for Japanese LVCSR

Fig. 7 Kaldi

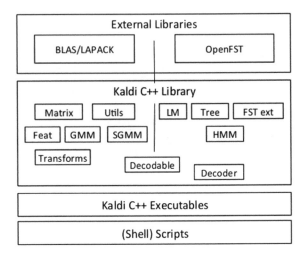

Fig. 8 Julius

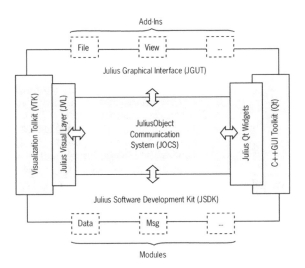

research from 1997, and the research continued in Japan's CSRS (Continuous Speech Recognition Consortium) from 2000 to 2003. The grammar based recognition parser called 'Julian' was integrated into Julius since rev.3.4.

The 'Julian' is a modified version of Julius that uses the manually designed DFA grammar as a language model. It could be used in constructing a small vocabulary speech command system or various speech interaction system tasks.

To run the Julius recognizer, it is needed a language model and acoustic model fitted for the language. Julius adopts the HTK ASCII typed acoustic model, pronunciation dictionary with the same type as HTK, and ARPA standard type word 3-gram language model. Although Julius is distributed only to Japanese models, the VoxForge project is doing the work of constructing an English acoustic model to be used with the Julius speech recognition engine. Recently, a new English ENVR-v5.4 open source speech model was released together with the Poland PLPL-v7.1 model in April 2018 thanks to the effort of Mozilla Foundation.

(4) *HTK*

HTK is a toolkit to process HMM. It is primarily aimed at speech recognition, but has been widely used in other pattern recognition applications using HMM including speech synthesis, character recognition and DNA sequencing. HTK developed for the first time by Machine Intelligence Laboratory (previously known as Speech Vision and Robotics Group) of Cambridge University Engineering Department (CUED) is now widely used by researchers who research HMM. HTK was developed originally in the Machine Intelligence Laboratory (old name: Speech Vision and Robotics Group) of Cambridge University Engineering Department (CUED) used for building the CUED's large vocabulary speech recognition system. In 1993, Entropic Research Laboratory Inc. acquired the HTK sales rights, and the HTK development was completely transferred to Entropic in 1995 when Entropy Cambridge Research Institute was established. HTK was sold by Entropic until 1999 when Microsoft bought Entropic. Microsoft now provides CUED again with the HTK license, and provides support so that CUED could redistribute HTK and provide development support through the HTK3 web site (Fig. 9).

Microsoft holds the copyright to the original HTK codes, but allows everyone to change the source code and contribute to including in HTK3.

For the current release, HTK version 3.4.1 is the stable release version, and HTK version 3.5 beta was launched as the latest release.

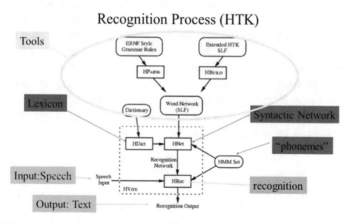

Fig. 9 HTK

Table 8 Open source speech recognition solutions

Solution name	Applied technology	License	Development language	Supported language
CMU Sphinx	HMM	BSD Style	Java	English
HTK	HMM Deep neural net.	HTK Specific License	C	English
Julius	HMM trigrams	BSD-Like	C	English, Japanese
Kaldi	Deep neural net.	Apache	C++	English

4   Conclusion

This study compared and analyzed commercial speech recognition software and speech recognition open source software. First, the features and functions of commercial speech recognition software such as Dragon Mobile SDK, Google Speech Recognition API, Siri, Yandex SpeechKit and Microsoft Speech API were analyzed, and as a comparison to them, CMU Sphinx, Kaldi, Julius and HTK among open source speech recognition software were analyzed.

Google is used the most among commercial speech recognition solutions. In particular, Google acquires many users in Korea in terms of supporting Korean language service. On the other hand, for open source speech recognition solutions, Korean speech data is not provided, so there is a difficulty that should obtain it separately (Table 8).

Of course, for Korean speech data, it is possible to collect through web crawling, but most speech files collected by crawling are difficult to use because they have many noise, and even if a clear speech file is collected, it should go through a process of transcribing into text. Despite these limitations, among the latest open source speech recognition solutions, HTK and Kaldi are rated as not inferior to commercial solutions also in the aspect of functions and performance, so they are being used in developing services in various fields. If supporting various languages such as Korean is given in the future, it is expected that open source speech recognition solutions could offer enough value as an alternative.

References

1. Kim, J.-H., Song, C.-W., Kim, J.-H., Chung, K.-Y., Rim, K.-W., Lee, J.-H.: Smart home personalization service based on context information using speech recognition. J. Korea Contents Assoc. **9**(11), 80–89 (2009)
2. http://www.segye.com/newsView/20170724003054 (2017)
3. http://news.hankyung.com/article/2017072469031 (2017)
4. Kim, S., Kim, J.-B.: A study on open source software business model based on value. Asia-Pac. J. Multimedia Serv. Convergent Art Humanit. Sociol. **7**(2), 237–244 (2017)

5. Sohn, H-j, Lee, M.-G., Seong, B-m, Kim, J-b: Collaboration tools of the Agile methodology impact on the OS project evaluation. Asia-Pac. J. Multimedia Serv. Convergent Art Humanit. Sociol 5(6), 9–18 (2015)
6. Telecommunications Technology Association (http://www.tta.or.kr/)
7. Park, H.S., Kim, S.W., Jin, M.-H., Yoo, C.D.: Current trend of speech recognition, based machine learning. Mag. IEIE. 18–27 (2014)

Implementation of Electronic Braille Document for Improved Web Accessibility

Ho-Sung Park, Yeong-Hwi Lee, Sam-Hyun Chun and Jong-Bae Kim

Abstract Visually impaired people tend to use printed text when their residual vision is high and braille when their residual vision is low. Braille is a medium that enables visually impaired people to see the world through texts, and they use braille to understand and express a variety of information required in daily life. Owing to developments in technology in recent years, visually impaired people can capture images of documents or objects using a camera mounted on their eyewear or using a smartphone camera and, after extracting texts and converting them into speech, they can interpret the information by listening to the audio content. As a result of substantial research and effort over a long period of time, significant technological progress has been made in enabling visually impaired people to access information. In addition to the development of technology, there are examples where the improvement in information accessibility for visually impaired people is due to improvements in policies and institutions. A typical example is accessibility to the Web, for which the most important means of information access is voice. However, to understand written texts, the most-frequently used learning method involves repeated reading. Hence, for more accurate understanding of the information, the means of "reading" is important, and for visually impaired people, "reading" refers to braille reading. In this study, first, the problems associated with the access to electronic documents that are provided by an information system through the Web or email are examined from the perspective of information access of visually impaired people. Second, a

H.-S. Park (✉)
Department IT Policy and Management, Soongsil University, Seoul, Korea
e-mail: hspark@atsoft.kr

Y.-H. Lee
ATSoft Co., Ltd., Seoul, Korea
e-mail: yhlee@atsoft.kr

S.-H. Chun
Department of Law, Soongsil University, Seoul 156-743, Korea
e-mail: shchun@ssu.ac.kr

J.-B. Kim
Startup Support Foundation, Soongsil University, Seoul, Korea
e-mail: kjb123@ssu.ac.kr

© Springer Nature Switzerland AG 2020
R. Lee (ed.), *Computational Science/Intelligence and Applied Informatics*,
Studies in Computational Intelligence 848,
https://doi.org/10.1007/978-3-030-25225-0_2

technique is implemented to convert data of a electronic document into braille in the information system server for enabling visually impaired people to access documents more accurately. Third, this study proposes a design method that can translate tabular information of civic service documents to braille by supplementing the disadvantages of conventional braille translation programs. Fourth, an institutional method is proposed for improvement in the efficiency of the aforementioned technique. It is hoped that the results of this study can be used to expand the electronic braille document infrastructure, which will enable visually impaired people to access a variety of information with increased accuracy. Based on this, we intend to provide a foundation for social participation.

Keywords Blind · Electronic braille document · Web accessibility · TTS · Reporting tool

1 Introduction

The main subject of the Davos Forum realized in January 2016 was "Mastering of the Fourth Industrial Revolution," and discussions were held on how to prepare for the big worldwide economic, cultural, and social changes. Representative information and communication technologies (ICT) of the fourth industrial revolution include artificial intelligence (AI), Internet of Things (IoT), big data, augmented reality, and Blockchain, which has become a focus of interest recently. Using object recognition, text recognition, and text-to-speech (TTS) conversion technologies, visually impaired persons can extract a text from an image captured with a camera or analyze the characteristics of an object stored in a database (DB), thereby hearing its name or text expressed in voice.

Starting with assistive devices such as the braille information device and screen reader, a variety of technologies and products such as the optical character reader (OCR) and DAISY player have been developed and used. As a result, the information accessibility of the visually impaired has improved greatly. Particularly, owing to the efforts made by the government to standardize the South Korean Web content accessibility guidelines and make them legally mandatory, visually impaired people can access Web pages more easily.

People visit Web sites that provide information regarding civic service applications, finance, education, the judicial system, and medical services for social participation and economic activities and acquire relevant information or issue the necessary civic service documents. Websites handle sensitive private information or contain information of legal value. To prevent leaking of private information, security programs, such as screen capture prevention, preview prevention, and keyboard security programs, are used. From the perspective of information access, these security programs lead to a problem whereby visually impaired people cannot use a screen reader to read the information of documents issued from information systems.

In particular, severely visually impaired people receive help from other people to review a variety of information such as civic service documents, bankbooks, notices, transaction details, bills, medical reports, and receipts. Here, another problem experienced by visually impaired people is the leakage of private information.

When a civic document is issued, if there is a system that can download a braille document containing identical information or output a braille document when the user wants, the visually impaired person will be able to read the corresponding document through braille without any help from other persons. Consequentially, he/she will be free from the privacy damage caused by serious leakage of private information.

This study points out to cases and problems of Web pages that are difficult to access and proposes a electronic braille document generation solution to provide braille documents from information systems. First, by using open source software, a braille conversion technique operating in an information system server is implemented; then, a method for improvement of the institutional efficiency is proposed so that the corresponding technique can be applied and used in the information system. Through this, it is expected that the electronic braille document generation solution will be adopted in conventionally operated information systems, thereby improving the information accessibility of visually impaired people and expanding their opportunities for social participation.

The remainder of this paper is organized as follows. Section 2 examines previous studies and relevant technologies related to braille such as assistive devices, Web accessibility guidelines, and open sources for information accessibility of visually impaired people. Section 3 describes the design of the proposed solution. Section 4 discusses the results of the solution implementation and Sect. 5 analyzes the experimental results. Lastly, Sect. 6 presents the implications of this study, the method for improvement of the efficiency, the limitations of this study, and a direction for the follow-up study.

2 Related Studies

The methods for information access of visually impaired people can be mainly classified into reading and listening. Of those, listening has been the most frequently used with the advancement of technology. As mentioned in Introduction, based on the progress of smart devices, cameras, and character decoding technologies, many devices have been released that take a photograph of text printed on paper using a camera, and then read it with voice after converting it to digital text through the OCR technology. However, from the perspective of learning for acquiring information, Kornell said that the learning method of repeated reading is commonly used to understand a written text [1]. In other words, reading is the best access method for understanding and memorizing information accurately. For individuals with severe visual impairment (who have no residual vision at all and have to rely on the tactile sense alone) reading refers to braille reading [2].

As the technology has advanced, many assistive devices that access information using voice have been developed. Nevertheless, from many aspects of daily life, work life, and self-esteem, along with information access, braille is a very important means of reading, which is confirmed by reports of many South Korean and overseas researchers including Kim, Park, Schroeder, Ryles, and Koenig [3–13].

Braille means more than a writing system to the visually impaired people. Braille lets the visually impaired people have self-confidence, independence, and equal rights [14]. Furthermore, visually impaired people who can read and write braille proficiently have a higher employment rate and higher self-esteem than those who do not know braille [15].

In every state of the USA, a braille law is enforced, stipulating that braille should be taught mandatorily to every child who needs it [16]. In other words, although visually impaired people have no difficulty in the spoken language activities of speaking and listening, they experience many difficulties due to the visual damage in the written language activities of reading and writing.

Recently, owing to the advancement of assistive devices for information and communication, braille information devices are frequently used by visually impaired people. These devices facilitate reading of braille using electronic braille files instead of papers. The use of a braille information device is preferred over braille paper because of the volume aspect and ease of document modification and sharing. However, considering the characteristics of the information provided by information systems, one of the main reasons for this preference is that a certain part or text string can be easily found. The inconvenience in information access does not just end as inconvenience. When dealing with bank functions, hospital prescriptions, or legal court issues, it goes beyond the inconvenience and leads to the exposure of private information or difficulty in keeping personal confidential information.

If various contracts, insurance policies, product manuals, and so on are provided in braille files so that visually impaired people can read them at any time, the visually impaired people will be no longer an alienated class in terms of using written texts in the digital society.

A electronic braille document refers to a document produced by converting a electronic document in a braille format of digital type. Braille translation programs have been developed for a very long time, and all the conventional braille translation programs translate texts into braille. However, to enable visually impaired people to understand the information in a document accurately, the information contained in the form of various tables should also be translated into braille.

3　Design of the Electronic Braille Document

3.1　Overview of the Solution

The basic function of the electronic braille document generation solution is that, interoperating with a reporting tool installed and used in the server, it converts data and form information into an electronic braille file. The reporting tool refers to a development tool that outputs result values extracted from the system in a report format using tables or graphs.

First, electronic documents, such as certificates, notices, confirmations, transaction statements, policies, and contracts, that are issued from information systems operated by public, educational, financial, and medical organizations are converted into braille files that can be accurately accessed by visually impaired people. Then, they are downloaded from the Web or received in an email and read through a braille information device.

The proposed electronic braille document generation solution operates in the information system server. Because it supports Web standards such as multi-OS and multi-browsers, does not use ActiveX, and does not require separate program installations, it can increase the convenience of use for visually impaired people.

From the overall system aspect, by designing the interoperation part with a currently used reporting tool, this study finds differentiating factors from conventional braille translation programs and proposes a method to improve the efficiency of the solution.

3.2　Design of the Solution

The automatic electronic braille document generation solution consists of three main modules for format and text of the electronic document, interoperation with the reporting tool, and braille file generation. More specifically, it consists of a format and text processing module for providing the tabular information of a form in braille; an API module for linking the reporting tool and data; and a module that finally generates a braille file.

After accessing the Web page of the information system, if a visually impaired person goes through a process of user authentication and requests access to a necessary document, the form and data of the corresponding document are rendered in the server and then a preview is displayed on a personal computer through the Internet.

In the preview screen, if the "electronic braille download" button is clicked, a braille file is generated in the server as shown in Fig. 1, and that file is downloaded. To resolve the inconvenience involved in the downloading process of a braille file on the Web for visually impaired people, the braille file can be sent automatically by email by simply clicking the document output button if the user is check-marked as a visually impaired person.

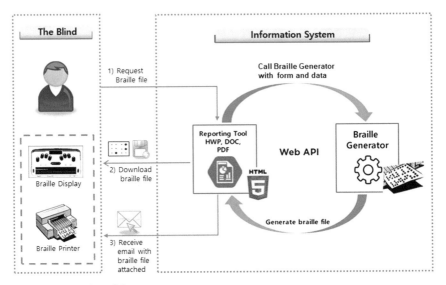

Fig. 1 Configuration of the system

3.2.1 Form and Text Processing Module

Given that the structure and method of expression are different between printed text and braille text, attributes such as document title, large heading, small heading, and table content should be specified in order to translate a regular printed document to a perfect braille document. In this study, these attributes are defined and when the form and data of the document have been processed according to its attributes, the resulting file is called a Braille Translator Markup Language (BTML) file. A BTML file refers to a file whereby the content and format of the original document are analyzed and the attributes required in braille translation and the information required for braille file conversion are recorded together with the original printed text content. The BTML file is finally converted to a braille file by using the braille translation module. That is, it can be seen as a file that keeps the required data before finally generating a braille file.

The BTML file of Fig. 2 consists of header, markup data for braille translation, and braille information regarding output. In the header, the version information of the file is recorded, and in the markup data for braille translation, page-change lines corresponding to the page units of the original document are recorded along with the original text information, in which the braille translation attributes have been added. Typical braille translation attributes include @@document name@@, @@page separation line @@, @@begin table @@, @@table separation line@@, and @@end table @@.

In the braille information regarding output, the braille file conversion information required for the output is recorded (e.g., number of lines, header, and footer).

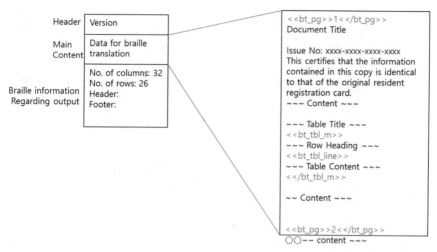

Fig. 2 Structure of a BTML file

The form and text extraction module extract the document name, page number, document content, table format in the document content, and table content from the electronic document, which has been generated by interoperating with the reporting tool. Then, it generates a Braille Text Object Model (BTOM). For the data excluding the tables, the field names and field values of the form are also inserted into the BTOM. For the table format and table data, the result values obtained by calling the table data processing module are additionally inserted into the BTOM.

Figure 3 classifies the table types and summarizes the rules regarding how to translate each attribute contained in the tables. The table data processing module finds a matching type in the table type DB by using the extracted table information and modifies the extracted table information according to the BTML rules. Then, it generates markup table data according to the table format.

In the table type DB, approximately 30 types of table formats used in regular civic certificates are defined along with the BTML rules based on them. If a table format that does not exist in the table type DB is found, it can be added in the DB.

3.2.2 Reporting Tool Interoperation Module

When a user requests issuance of certain document on the Web, a reporting server existing in a Web Application Server (WAS) is called. The reporting server fetches the corresponding document form and data and after making an electronic document, downloads it to the user's PC through the Web.

In this process, by generating a braille file after calling the braille conversion library for visually impaired people, a download service can be provided together with the original digital document. Furthermore, when the original copy of the document

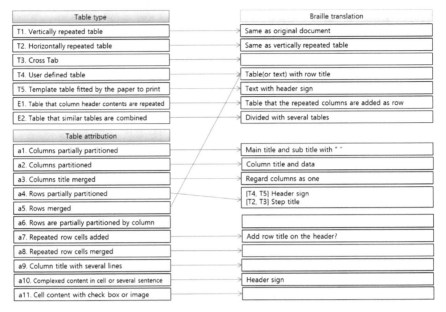

Table type	Braille translation
T1. Vertically repeated table	Same as original document
T2. Horizontally repeated table	Same as vertically repeated table
T3. Cross Tab	
T4. User defined table	Table(or text) with row title
T5. Template table fitted by the paper to print	Text with header sign
E1. Table that column header contents are repeated	Table that the repeated columns are added as row
E2. Table that similar tables are combined	Divided with several tables

Table attribution	
a1. Columns partially partitioned	Main title and sub title with " "
a2. Columns partitioned	Column title and data
a3. Columns title merged	Regard columns as one
a4. Rows partially partitioned	[T4, T5] Header sign
a5. Rows merged	[T2, T3] Step title
a6. Rows are partially partitioned by column	
a7. Repeated row cells added	Add row title on the header?
a8. Repeated row cells merged	
a9. Column title with several lines	
a10. Complexed content in cell or several sentence	Header sign
a11. Cell content with check box or image	

Fig. 3 Braille translation rules for tables by type

is converted to a PDF file by the reporting tool and sent to the user by email, the braille file can then be attached along with the PDF file and sent out.

The process of interoperation with the reporting tool is shown in Fig. 4.

Figure 5 shows a design of the API (Application Programming Interface) method between the reporting tool and the electronic braille generation solution. In other

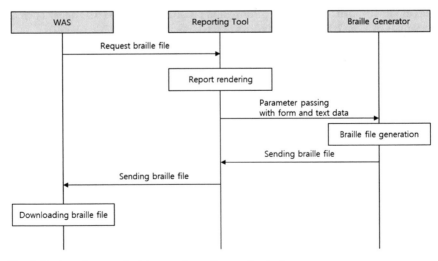

Fig. 4 Design of process for interoperation with reporting tool

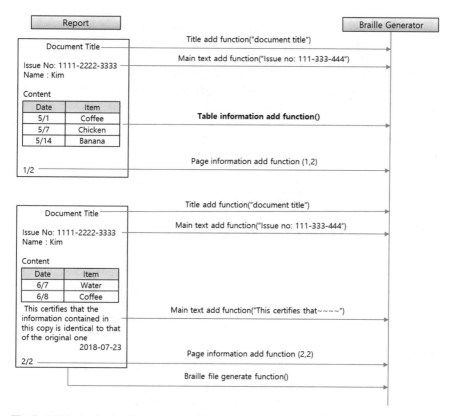

Fig. 5 API design for braille generation of document content using the reporting tool

words, it shows an API design for data communication based on prearranged functions to perform braille translation of the document structure, text, and table contained in a certain document.

3.2.3 Braille File Generation Module

The BTML file generation module generates a BTML file by using the document information registered in the BTOM. The document title in the BTOM is saved in the BTML file by adding six blank spaces in front. Moreover, a table in the BTOM is saved in the BTML file by adding "@@begin table @@" in front of the table content. In order to convert the BTML file into an output braille file, the number of lines, header, and footers should also be recorded.

The BTML interpreter reads the BTML one line at a time, and if a braille translation code that shows a braille translation attribute is read, it is replaced with a predefined braille, and if regular text character information is read, it is replaced with translated braille through the developed braille translation engine. In this way, a braille file is generated for the braille device. If an output braille file is needed, it can be generated using the braille file generation module and the braille information regarding output in the BTML.

The braille file generation module retrieves the line and column information stored in the braille information of the BTML file pertaining to output. Then, it modifies the maximum number of spaces per line and the maximum number of lines per page in the generated braille file, and after braille-translating and adding the footer and page number at the bottom of each page, it generates a final output braille file.

4 Implementation of the Solution

4.1 Implementation Environment

As for the implementation result of the electronic braille document generation solution, the size of the finally generated braille file is evaluated, thereby evaluating the effect on the accessibility of visually impaired people and the server performance. To confirm that the solution can run on an information system server, the library developed using the open-source software is examined to verify whether it operates normally on the server.

As shown in Fig. 6, this study implements the following to braille-translate the information of a table: data are prepared in advance for braille translation of form and text, and when they are inputted in the input window of the Web browser, a braille file is generated as a result. The actual implementation of interoperation with the reporting tool that is part of the design is not included in this study because it requires collaboration with the corresponding company.

As shown in Fig. 6, the implementation process in this study is as follows: an electronic braille file is generated from the information system server, and by checking the size of the generated file, it is investigated whether there is a problem when

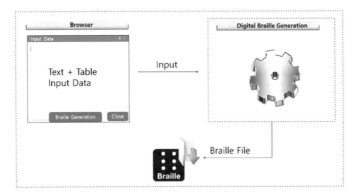

Fig. 6 Implementation process

downloading it from the Web or sending it in an email. Prior to this, by using open source software, an open-source test was conducted, and the Korean braille rules were applied through the simple example shown in Fig. 7.

The open source Liblouis braille translation software contains a Korean braille translation table. Therefore, a simple braille translation sample test was possible through the open source software, and for this purpose, the revised Korean braille rules were reflected.

Figure 8 shows a partial sample of the braille translation table in Korean syllable units contained in the open source software. In braille, one character is completed with six points, and here, numbers indicate the position of the respective point.

```
int
main(int argc, char **argv) {
        int optc;

        set_program_name(argv[0]);

        while ((optc = getopt_long(argc, argv, "hvfb", longopts, NULL)) != -1) {
                switch (optc) {
                /* --help and --version exit immediately, per GNU coding
standards. */
                        case 'v':
                                version_etc(
                                                stdout, program_name, PACKAGE_NAME,
VERSION, AUTHORS, (char *)NULL);
                                exit(EXIT_SUCCESS);
                                break;
                        case 'h':
                                print_help();
                                exit(EXIT_SUCCESS);
                                break;
                        case 'f':
                                forward_flag = 1;
                                break;
                        case 'b':
                                backward_flag = 1;
                                break;
                        default:
                                fprintf(stderr, "Try `%s --help' for more information.\n",
program_name);
                                exit(EXIT_FAILURE);
                                break;
                }
        }
```

Fig. 7 Open-source program used for testing

```
# There are seven Korean conscenant dots which may be confused with numberals.
# In this case, a speace character (dot 0) is inserted between end of number(s) and Korean
charaacters.
after digit always 나 0-14-126
after digit always 냬 0-14-126-1
after digit always 냭 0-14-126-1-1
after digit always 낛 0-14-126-1-3
after digit always 냐 0-14-126-25
after digit always 냣 0-14-126-25-13
after digit always 냜 0-14-126-25-356
after digit always 넏 0-14-126-35
after digit always 널 0-14-126-2
after digit always 넑 0-14-126-2-1
after digit always 넒 0-14-126-2-26
after digit always 넓 0-14-126-2-12
after digit always 넋 0-14-126-2-3
after digit always 넔 0-14-126-2-236
after digit always 넖 0-14-126-2-256
after digit always 넜 0-14-126-2-356
after digit always 넍 0-14-126-26
after digit always 넌 0-14-126-12
after digit always 넟 0-14-126-12-3
after digit always 넛 0-14-126-3
after digit always 넜 0-14-126-34
after digit always 넣 0-14-126-2356
after digit always 넟 0-14-126-13
after digit always 넞 0-14-126-23
after digit always 넭 0-14-126-235
after digit always 넫 0-14-126-236
after digit always 넡 0-14-126-256
after digit always 넝 0-14-126-356
after digit always 녀 0-14-234-1
after digit always 녂 0-14-234-1-1
```

Fig. 8 An example of Korean table included in the open source software

The sample documents used for the final braille generation in this study included a resident registration certificate, certificate of eligibility for the National Health Insurance Service, certificate for National Pension subscription, receipt for earned income tax withholding, and transaction statement through Internet banking, which are the most widely used among the civic service forms. These are the typical forms that are most used, and because they consist of tables, they provide opportunities for investigating the needs of visually impaired people for understanding the Table 1.

Table 1 Hardware implementation environment

Category	Model and performance	Remark
WEB	IIS 7.0	
CPU	Intel Core2 3.0 GHz	
OS	Windows 7 Professional K	64 bit
Memory	2 GB	
HDD	70 GB	

4.2 Braille File Generation

This subsection examines if the module using the open source software operates normally on the server, and a braille file is generated as a result. As for the sample documents used for generation of braille files in this study, a total of 204 types were used, including 50 types of bank transaction statements. Figure 9 shows the image of a sample document (bank transaction statement) converted into braille.

5 Analysis of Experimental Results

In this study, the proposed electronic braille document generation solution was developed using open source software, and braille files were generated by running the solution on a Windows server. As a result, it was confirmed that the size of the generated braille file did not affect the performance of the conventional server. Table 2 summarizes the file types and sizes of Fig. 9 in an easy-to-read format.

In Table 2, the receipt for earned income tax withholding, which has the largest amount of original content, has a braille file size of 6 KB, and in most cases, the size of the braille file is less than 10% of the PDF file size. This confirms that the proposed solution does not cause severe load on the server and there is no problem in distributing electronic braille document files via download or email.

Table 3 compares the proposed solution with conventional braille translation programs. Among similar programs, A and B are programs that convert texts of HWP or DOC documents into braille. However, a runs on a Windows server, and provides a service that shows the braille translation results on the Web.

Both A and B facilitate braille translation of text only, but the proposed solution provides a braille translation method for tables. Furthermore, a sample file of a civic service document is generated in the reporting tool, and after confirming it in the viewer, the braille translation result is provided as data. This demonstrates the extendibility of the proposed solution in the sense that it can interoperate with the reporting tool, which is the most frequently used to output civic service documents.

Among the factors for running the solution on the information system servers, testing in diverse OS environments was neither implemented nor verified in this

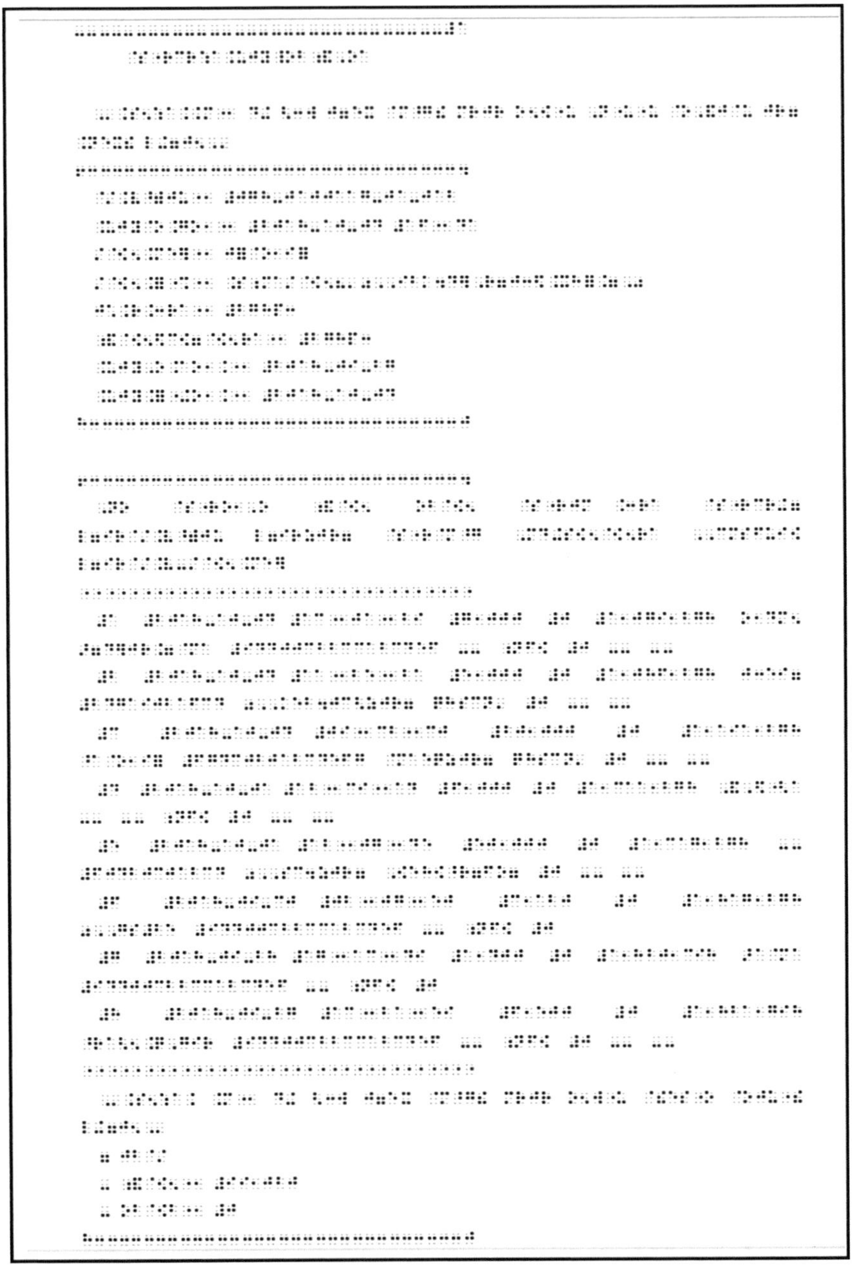

Fig. 9 Braille of bank transaction statement

Table 2 Comparison with original document size (unit: KB)

File type	Resident registration certificate	Certificate of eligibility for national health insurance service	Certificate for national pension subscription	Receipt for earned income tax withholding	Bank transaction statement
HWP size	19	42	52	41	50
PDF size	30	44	36	87	63
BRL size	3	1	2	6	2

Table 3 Comparison with conventional braille translation programs

Category	Proposed solution	A	B
Braille file	O	O	O
Operate on information system server	O	O	X
Interoperation format	Reporting tool, PDF, HWP, DOC, WedGrid	HWP DOC TXT	HWP DOC TXT
Server OS	Windows, Unix, Linux, HP-UX, IBM AIX, Solaris	Windows	N/A
Braille translation of table	O	X	X

study, although it was mentioned. Additional implementation and verification are required for various OS environments beside Windows, such as Linux and UNIX.

6 Conclusion

It was confirmed earlier that the use of braille is helpful for the social participation and economic activity of visually impaired people. From this perspective, important implications provided by this study are as follows.

First, electronic braille was mentioned as an alternative means for improvement of the Web accessibility of visually impaired people, and user convenience can be provided by installing it on information system servers. To this end, braille generation was implemented using open source software, and its comparative advantage over the conventional braille translation programs was confirmed.

Second, a design was proposed for implementation of an API that can convert form and data information into an electronic braille file by interoperating with the reporting tool. Through this design, the possibility that the information of tables required for accurate information access of visually impaired people could also be provided in braille was proved.

Third, the braille file was proposed as an alternative means that can improve the accessibility problem of Web pages when they cannot be accessed by a screen reader because of a screen-capture prevention program or when preview is not provided in the Web service process of electronic documents that have sensitive private information and legal force.

Fourth, considering that a technical method for braille-translating not only simple text but also information of tables was provided, it will be helpful in the development of techniques for braille translation of information in tables contained in conventional documents of HWP or DOC format. Therefore, in the future, to improve the efficiency, it is necessary to analyze many electronic documents that are distributed in PDF format and uploaded as HWP and DOC documents, and to develop an interface to process these formats.

Lastly, it was confirmed that a policy for electronic braille should be investigated as an alternative means to be included in braille laws, laws related to disability discrimination and protection of rights, and Web content accessibility guidelines.

References

1. Fellenius, K.: Swedish 9-year-old readers with visual impairment; a heterogeneous group. J. Vis. Impairment Blindness **93**, 370–380 (1999)
2. Fountain, J.: Building the Virtual State: Information Technology and Institutional Change. Brookings Institution Press, Washington D.C.*, pp. 88–98 (2001)
3. Ryles, R.: The impact of braille reading skills on employment, income, education, and reading habits. J. Vis. Impairment Blindness **90**(3), 211–215 (1996)
4. Schroeder, F.: A step toward equality: Cane travel for the young blind child. Future Reflections **8** (1989)
5. Sugano, A., Ohta, M., Oda, T., Miura, K., Goto, S.: eBraille: a web-based translation program for Japanese text to braille. Internet Res. **20**(5), 582–592 (2010)
6. Kim, M.-S., Choi, J.-H.: The Fourth Industrial Revolution and Understanding of Industrial IoT and Industrial Internet. KISDI New Report Vol. 28 No. 12, Vol. 626, Institute of Information and Telecommunication Policy, pp. 20–26 (2016)
7. Kim, Y.-Il., Lee, T.-h.: An investigation of the perspectives and current status on the use of braille among the individuals with visual impairments. Korean J. Vis. Impairment **31**(3), 157–177 (2015)
8. Kim, Y.-Il, Moon, H.: An examination of the current status and needs of braille literacy education for individuals with adventitiously visual impairments. Korean J. Vis. Impairment **33**(1), 1–29 (2017)
9. Park, S.-S.: Braille and Information Access in the Digital Age. 2018 Braille Anniversary Seminar Presentation, pp. 13–19 (2018)
10. Park, J.H., Oh, C.W.: Analysis on braille understanding and braille application condition of visually impaired people. Korean J. Vis. Impairment **27**(4), 135–157 (2011)
11. Lee, C.-W.: A Study on the Effect of Institutions on the Judiciary e-Government promotion. Seoul National University, The Graduate School Master Thesis (2018)
12. Hyun, J.-H., Kim, S.-i.: The actual condition and improvement methods of web accessibility and usability of Korea government department. KADO Issue Report No. 31(vol. 3 No. 7), National Information Society Agency (2006)
13. Hwang, S.-M.: A Study on Reading in Paperless and Paper Braille of the Students with Visual Impairment. Department of Special Education Graduate School of Education, KongJu National University (2006)

14. Greaney, J., Reason, R.: Phonological processing in braille. Dyslexia **5**, 215–226 (1999)
15. Koenig, A.J., Holbrook, M.C.: Learning media assessment. TSBVI, Austin (1995)
16. Kornell, N., Bjork, R.A.: The promise and Perils of self-regulated—58—study. Psychon. Bull. Rev. **14**(2), 219–224 (2007)

A Reliable Method to Keep Up-to-Date Rights Management Information of Public Domain Images Based on Deep Learning

Youngmo Kim, Byeongchan Park and Seok-Yoon Kim

Abstract Most public domain images are exposed to copyright infringement due to different Rights Management Information (RMI) expression system and/or out-of-date RMI. To solve this problem, this paper proposes a reliable method of integrating the RMI representation system and updating the RMI with the up-to-date information based on the most reliable data among information collected from each site through a comparative search technique for public domain images based on deep learning. Through the RMI updating method proposed in this paper, public domain image sites will be able to minimize copyright infringement, since they can provide the most up-to-date RMI and may transform different RMI representation systems into a unified system.

Keywords Public domain image · Rights management information · Deep learning · Weighted scoring model · Search engine

1 Introduction

The usage of public domain images has recently been versatile among individuals and businesses. In particular, single-person companies, developers and designers such as individuals and SOHO businesses are showing great interests in free-use works that are freely available without copyright concerns [1]. These images are available on public domain image service platforms such as the Gong-U Madang (Korea), Flickr, Europeana and Pixabay, and the total number of serviced public domain images may reach more than 1 billion. Despite the growing demand for public domain images, disputes over copyrights have arisen due to the usage without

Y. Kim · B. Park · S.-Y. Kim (✉)
Department of Computer Science and Engineering, Soongsil University, Seoul, Republic of Korea
e-mail: ksy@ssu.ac.kr

Y. Kim
e-mail: ymkim828@ssu.ac.kr

B. Park
e-mail: pbc866@ssu.ac.kr

© Springer Nature Switzerland AG 2020
R. Lee (ed.), *Computational Science/Intelligence and Applied Informatics*,
Studies in Computational Intelligence 848,
https://doi.org/10.1007/978-3-030-25225-0_3

a thorough understanding of copyrights [2, 3]. These problems arise from the out-of-date RMI and/or inconsistent RMI representation systems among different public domain sites. To solve these problems, studies have been conducted to integrate different RMI presentation systems among sites and to provide the most up-to-date RMI.

However, although users confirmed that duplicate images were discovered at two public domain sites when they searched for an image, a delicate issue may still arise if the right holder of that image partially updated his RMI, that is, if he updated the RMI only in one site. To solve this problem, this paper integrates the RMI representation system and presents a method for finding duplicate public domain images that are registered at different sites through a comparative search method based on deep learning. In addition, an updating method is proposed to keep up-to-date RMI for the collected images.

The composition of this paper is as follows. After this introduction, Sect. 2 surveys the deep learning technique and image feature point detection and matching algorithm. Section 3 proposes an integrated RMI representation system among public domain sites, a search method for duplicate images based on deep-learning, and an updating method to keep the most up-to-date RMI. Section 4 shows an experimental results and performance evaluation of the proposed method and conclusion is given in Sect. 5.

2 Related Research

2.1 CNN (Convolutional Neural Network)

The CNN is one of the fundamental architectures in the field of image recognition and its basic concept is illustrated in Fig. 1, where the convolutional layer is added to the pooling layer and is different from the fully connected layer shown in Fig. 2. Figure 2 shows that the 'Affine-RLU' configuration can be used in layers close to the output, and the 'Affine-Softmax' combination is used in the last output layer [2]. In CNN, input/output data of a composite product layer is called a Feature Map, input data of a composite product layer is called an Input Feature Map, and output data is called an Output Feature Map.

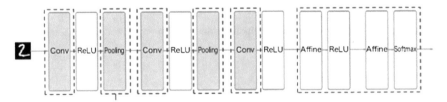

Fig. 1 Network with CNN

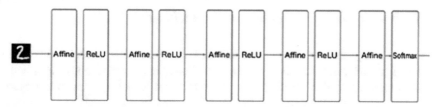

Fig. 2 Network consisting of a complete connection layer (affine layer)

2.2 YOLO (You-Only-Look-Once)

The YOLO algorithm is one of the deep learning-based object detection algorithms, and solves bouncing box detection and class classification defining them as one regression problem, unlike algorithms such as Faster R-CNN and Mask R-CNN using the Region Proposal method. Compared to the existing methods, YOLO resulted in similar detection results, but yielded excellent performance improvements in terms of elapsed time because the region proposal time is eliminated. The YOLO algorithm is equally divided into $S \times S$ grid cells with randomly set input images and predicts B bounding boxes for each grid cell. The information to predict the bounding box is predicted using the x coordinate, y coordinate, width, height, information, and probability of being a class. Of the projected B-bounding boxes, the context value defines the highest value as the class in the bounding box, and if the value of the contrast value exceeds the predetermined threshold, it is determined to be a valid bounding box [4, 5] (Fig. 3).

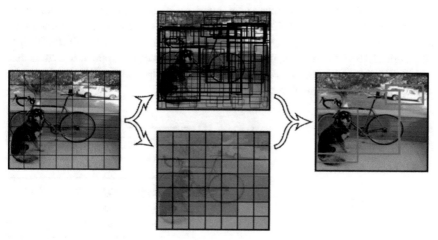

Fig. 3 YOLO action process

2.3 Feature Point Extraction and Matching

The matching process of images can be divided into 'Feature Detection' that extracts unique feature information and related information of images, 'Oriental Calculation' and 'Feature Extraction' processes for extracting and recording rotation angles as they change around the extracted feature points, and 'Feature Matching' process, which is the process for comparing differences between feature vectors.

In many studies where feature extraction and matching are performed, algorithms such as SIFT, SURF and ORB are typically used. Each algorithm includes an image matching process as illustrated in Fig. 4.

SIFT (Scale-Invariant Feature Transform) was completed in 2004 after Professor David G. Lowe began to study size and rotational immutable area characteristics in 1997. This algorithm was suggested by Harris Corner so that the cornering is maximized not only within the image but also on the scale axis, based on the Differential of Gaussian (DoG) to solve the sensitive issue of image change. It is essentially the feature which expresses the direction of local brightness change and the degree of rapid change around feature points, and is robust in size, shape, and direction (rotational) change, and has excellent distinction.

Speed-Up Robot Feature (SURF) is the method proposed by Herbert Bay in 2008 to allow for the extraction of tough features while improving speed over SIFT algorithms. SURF is an algorithm that utilizes SIFT to find interesting points and region using integral images to improve processing speed at each step of calculating Keypoint and Descriptor, and uses fast-Hessian detection to calculate interesting points, which is the area of image (the sum of the brightness). It is possible to extract tough features in blur or size deformation/rotation of images.

The ORB algorithm is an image characteristic detection algorithm developed by OpenCV LAB to replace SIFT and SURF, which are difficult to use commercially due to problems with licenses, with the algorithms applied mixed with FAST algorithm and BRIEF. In addition, pyramids of various scales are applied to extract size and immovable image characteristics, and similarly adjusted BRIEF algorithms are applied to extract rotational invariant properties [6–8].

Fig. 4 Image matching process

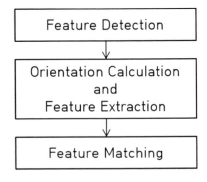

3 Public Domain Image Comparison and RMI Update Method Based on Deep-Learning

Figure 5 shows the architecture of the proposed method of comparing and updating the deep learning-based public domain images.

The proposed system first collects the required images and metadata from the public domain site using Web Crawler technology and stores them in the HDFS-based Image Processing DB.

Then the system extract and label objects for the public domain images extracted through the YOLO module to store keywords and use the Feature Point Extraction module to check for duplicate images collected from public domain sites.

Since the collected images are large in volume, the system extracts feature points through the ORB algorithm only for the images with the same keyword extracted from the YOLO module and try to find duplicate images in the Matching module.

If such duplicate images are present, RMI information is updated based on reliable results using the RMI update technique.

3.1 RMI Integration

The RMI expression system of each site is integrated into one to reduce confusion among users and facilitate the management of an integrated RMI system due to different rights management information expressions among sites.

RMI is the expression method of RMI on the sites of shared works, and only works that are freely available, such as expired works and orphaned works, are incorporated into free use works, as in Table 1.

Tables 2 and 3 show the integrated RMI classification of 'Gong-U Madang' (Korea) and Flickr and Europeana, respectively, and such RMIs that there are no

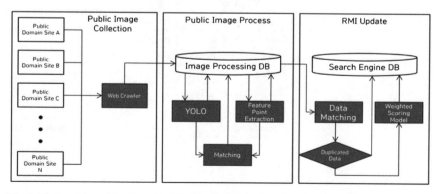

Fig. 5 RMI update architecture of public domain images based on deep learning

Table 1 RMI database scheme

RMI_code	RMI_eng	Explanation
rmi_01	CC-BY	Share—copy and redistribute the material in any medium or format Adapt—remix, transform, and build upon the material for any purpose, even commercially
rmi_02	CC-BY-NC	Share—copy and redistribute the material in any medium or format Adapt—remix, transform, and build upon the material
rmi_03	CC-BY-ND	Share—copy and redistribute the material in any medium or format for any purpose, even commercially
rmi_04	CC-BY-SA	Share—copy and redistribute the material in any medium or format Adapt—remix, transform, and build upon the material for any purpose, even commercially
rmi_05	CC-BY-NC-SA	Share—copy and redistribute the material in any medium or format Adapt—remix, transform, and build upon the material
rmi_06	CC-BY-NC-ND	Share—copy and redistribute the material in any medium or format
rmi_07	FREE USE	The author can give up his rights or use it without any restriction over the copyright term

Table 2 Integrated RMI schema of 'Gong-U Madang'

Gong-U Madang RMI	Integrated RMI	Remarks
CC-BY	rmi_01	
CC-BY-NC	rmi_02	
CC-BY-ND	rmi_03	
CC-BY-SA	rmi_04	
CC-BY-NC-SA	rmi_05	
CC-BY-NC-ND	rmi_06	
Expired Assets	rmi_07	
Donation (Free use)	rmi_07	
Donation (Permission)	rmi_07	
Gong-Gongnuli 1 Type	X	Not image format
Gong-Gongnuli 2 Type	X	Not image format
Gong-Gongnuli 3 Type	X	Not image format
Gong-Gongnuli 4 Type	X	Not image format

Table 3 Integrated RMI Schema of Flickr and Europeana

Flickr RMI	Integrated RMI	Remarks
CC-BY	rmi_01	
CC-BY-NC	rmi_02	
CC-BY-ND	rmi_03	
CC-BY-SA	rmi_04	
CC-BY-NC-SA	rmi_05	
CC-BY-NC-ND	rmi_06	
No known copyright restrictions	rmi_07	
Public domain	rmi_07	
All rights reserved	X	Copyright protected
U.S. government works	X	Copyright protected

restrictions in using the public domain works are integrated into 'rmi_07', which means 'FREE USE'.

In addition, some images are excluded from 'FREE USE' in cases that they are not stored in the image format, or they are not freely available to users for general purposes, or their copyright information is not clear so that the copyright dispute might occur in the future.

3.2 Image Comparison Based on Deep Learning

For images collected through Web Crawler, object recognition is performed on images based on deep learning for fast comparison of duplicate images. Deep learning system uses the YOLO algorithm to recognize objects, label and classify objects.

In order to extract objects from the collected public domain images, the objects were first studied, where the data set was based on PASCAL VOC 2007, and the object classification are shown in Table 4.

You can see that objects are extracted, as shown in Fig. 6. Objects separated in Fig. 6 could be identified as four horses.

Table 4 Examples of objects

Division	Object
Person	Person
Animal	Bird, cat, cow, dog, horse, sheep
Vehicle	Aeroplane, bicycle, boat, bus, car, motorbike, train
Indoor	Bottle, chair, dining table, potted plant, sofa, tv/monitor

Fig. 6 Object detection using YOLO

Finding duplicate images at each site requires information about objects classified by the YOLO algorithm. Because of the large number of public domain images, comparing them to all images takes a long time, so that the YOLO module sorts objects for all public domain images collected through Web Crawler and checks for duplicate images through ORB algorithms based on classified objects [9, 10].

3.3 RMI Renewal Model

If matching data is detected as duplicate data at multiple sites, the problem arises as to which sites to determine the most accurate data, and as a workaround, RMI renewal models are designed to be updated based on the most reliable data. By using RMI update models that determine data reliability, this paper proposes to evaluate items such as Table 5.

The evaluation items in Table 5 is classified into two categories: one that measures the reliability of a site, called Site Reliability, and the other that measures the reliability of posts within the site, called Posts Reliability. Site Reliability depends on the operating organization of site, the number of visits to the site and the number of

Table 5 Evaluation items of weighted scoring model

Site reliability	Posts reliability
Number of recent visits	Number of views
Number of recent posts registered	Download count
Total posts registration	Meta information count
Upload type	Update date

Table 6 Rating by weighted scoring model

Site reliability		Posts reliability	
Category	Weight (%)	Category	Weight (%)
Number of recent visits (nrv)	30	Number of Views (nv)	10
Number of recent posts registered (nrp)	20	Download Count (dc)	25
Total posts registration (tpr)	20	Meta Information Count(mic)	15
Upload type (ut)	30	Update Date (date)	50

postings registered within the last month, the total number of postings registered and the method of registration. Assessment items such as total number of posts registered and number of visits to the site within the last month are normalized between the maximum and minimum.

Finally, there are two ways to upload a personal work: direct registration to a site such as in Flickr, and the registration with the approval of an administrator. If an application for a change in the RMI was approved by the administrator, it could be somewhat delayed due to the administrator's processing time, but the RMI in the direct registration to the site can be modified immediately at all times, so a higher weight value was assigned.

Assessment tables were prepared with different weights on items such as the number of meta-information about a work. For the number of views and the number of downloads, the scores were distributed by comparing the postings of duplicate works sites to each other and depending on how much meta-information was provided about the works.

If the update date is the most recent, we put the greatest weight on the reliability score. Each item in the two evaluation tables has a different weight depending on its importance, and the weight on each item is given by Table 6.

Based on Table 6, the total reliability score, Reliability Score, is given by the expression:

$$Site\ Reliability = 30\% * nrv + 20\% * nrp + 20\% * tpr + 30\% * ut$$
$$Posts\ Reliability = 10\% * nv + 25\% * dc + 15\% * mic + 50\% * date$$

4 Experimental Results and Performance Evaluation

In this paper, the experimental performance evaluation was performed to integrate the RMI expression system and to find the efficient method through the deep learning based comparative search technique. In the experiments for 6000 images, the sample images with keywords are compared against images with the same keywords at 4

Table 7 Performance results of comparative search on image feature points

	Processing speed	Accuracy
ORB similarity detection	8 min 3 s	90%

different sites(denoted as site A though site D) and the performance results of this comparative search is given in Table 7.

The same image shown in Fig. 7 was detected on site A and site D when the image was renewed. In such cases, the actual data for each site are first collected and scores are calculated for each site to obtain reliability score. Table 8 shows the values for evaluation items at each site, and Table 9 illustrates the site-specific reliability scores once the weights are counted.

Fig. 7 Example of duplicated images at different sites

Table 8 Results of evaluation items at public domain sites

	Site A	Site B	Site C	Site D
Number of recent visits	9876	17,554	2350	6286
Number of recent posts registered	1241	3211	890	638
Total number of posts registration	14,855,462	485,321.540	9.345.621	13,669,748
Uploading method	Admin approval	Direct registration	Admin approval	Direct registration

Table 9 Reliability scores calculated at public domain sites

	Site A	Site B	Site C	Site D
Number of recent visits (weight 30%)	8.43	15.00	2.00	5.37
Number of recent posts registered (weight 20%)	3.86	10.00	2.77	1.98
Total number of posts registration (weight 20%)	0.30	10.00	0.19	0.28
Uploading method (weight 30%)	0.00	15.00	0.00	15.00
Site reliability score	12.59	50.00	4.96	22.63

The scores for each item in Table 9 were calculated up to 50 points. The last item, Uploading Method, in Table 9 means that the author directly registers/modifies his work, and adds 15 points to sites B and D.

Table 10 shows the values for postings on sites A and D where duplicate images were detected, and Table 11 illustrates the Posts Reliability scores after portraying the Site Reliability scores at sites. The score calculation formula for each assessment item is the same expression above that was used to obtain the site reliability, and only the site with the latest update date has been assigned additional points.

Now we can compare the reliability of the two sites using both Site Reliability Score calculated earlier and Posts Reliability score. At this time, the weight factors are given as 40% for Site Reliability and 60% for Posts Reliability. Total Reliability for site A is 19.58 and Total Reliability for site D is 37.31, as shown in Table 12, which claims that the data for site D is considered more reliable and the updating operation is executed based on site D data.

Therefore, when an public domain image is uploaded to a different sites with different RMI, it can be verified that the RMI is updated from 'rmi_04' to 'rmi_02' according to the reliability criteria of the postings, as shown in Fig. 8.

Table 10 Results of evaluation items for duplicate image articles

	Site A	Site D
Number of views	145	170
Download count	65	50
Meta information count	8	8
Update date	2016-01-24	2017-06-18

Table 11 Posts reliability scores calculated for duplicate image articles

	Site A	Site D
Number of views (weight 10%)	4.26	5.00
Download count (weight 25%)	12.5	9.61
Meta information count (weight 15%)	7.50	7.50
Update date (weight 50%)	0	25.00
Posts reliability score	24.26	47.11

Table 12 Reliability comparison of site A and D

	Site A	Site D
Site reliability (weight 40%)	5.03	9.05
Posts reliability (weight 60%)	14.55	28.26
Total reliability	19.58	37.31

seq	image_src	title	creater	right_code
655223	C:/Users/dukih/Desktoo/serarch enoine/imaoe/...	Reflejo en barro (Monument Valley)	Juan Luis Díaz	rmi 04

kor_rmi		source_code	org_url	reg_date	last_update_date
저작자표시-동일조건변경허락	FK		https://www.flickr.com/photos/nufus/30893922...	2018-01-24	2018-06-22

seq	image_src	title	creater	right_code
655223	C:/Users/dukih/Desktoo/serarch enoine/imaoe/...	Reflejo en barro (Monument Valley)	Juan Luis Díaz	rmi 02

kor_rmi		source_code	org_url	reg_date	last_update_date
저작자표시-비영리	FK		https://www.flickr.com/photos/nufus/30893922...	2018-01-24	2018-11-23

Fig. 8 RMI update results based on weighted scoring model

5 Conclusion

In this paper, we proposed an architecture that updates RMI using the most up-to-date information for public domain images so that users can find correct images and their RMI in the search engine. For the performance evaluation of comparative search using the proposed method, the experiments were performed for 6000 images and the results identified some duplicate images by comparing the only images with the same keyword for fast search speed, based on keywords for objects extracted through features point comparison based on deep learning. In addition, the image search engine could provide the original image and the latest RMI even if the user retrieved the revised image through the updating of RMI. Therefore, it is expected that the proposed method will improve the accuracy of the image search engine by updating the RMI based on reliable public domain sites and postings, and will help users to find the correct works by providing users with accurate original works and their RMIs using deep learning even if users happen to find some modified works.

Further research will require the classification of data sets for improving image feature learning speed, the application of improved deep learning algorithms, and the ongoing study of selection of evaluation items to provide more reliable data to keep up-to-date RMI, as well as the study on recommended images through the image object keyword based on deep learning.

Acknowledgements This research project supported by Ministry of Culture, Sport and Tourism (MCST) and Korea Copyright Commission in 2019 (2017-SHARE-9500).

References

1. Seo, E., Kim, I., Hong, D., Kim, Y., Kim, S.Y.: Integrated multilayer metadata based on intellectual information technology for customization services of public domain images with different usage permission of license. 28th J. Theor. Appl. Inf. Technol. **96**(4), 1048–1058 (2018)
2. Kim, M.J., Park, B.: A study on big data application using scoring model. J. Actuar. Sci. **7**(2), 3–22 (2015)
3. Ren, S., He, K., Girshick, R., Sun, J.: Faster R-CNN: towards real-time object detection with region proposal networks. IEEE Trans. Pattern Anal. Mach. Intell. **29**(6), 1137–1149 (2017)
4. bskvision, http://bskyvision.com/21
5. Krizhevsky, A., Sutskever, I., Hinton, G.E.: ImageNet classification with deep convolutional neural networks. In: Proceedings of the NIPS'12 Proceedings of the 25th International Conference on Neural Information Processing Systems, vol 1, pp. 1097–1105 (2010)
6. HackerFactor, http://www.hackerfactor.com/
7. Hikodukue, K.: Python Ni Yoru Scraping & Kikaigakushu Kaihatsu Technique. pp. 301–320. Japan-Press (2016)
8. Lowe, D.G.: Object recognition from local scale-invariant features. In: Proceedings of the Seventh IEEE International Conference on Computer Vision, vol. 2, p. 1150 (1999 Sept)
9. Goki, S.: Deep Learning from Scratch, pp. 227–235. Japan-Press (2016)
10. Ji, S., Kim, H., Cho, J.: Estimation of non-trained category in image classification model based on deep-learning. J. Inst. Control Robot. Syst. **24**(9), 793–801 (2018)

Applying GA as Autonomous Landing Methodology to a Computer-Simulated UAV

Changhee Han

Abstract The ultimate purpose of this study is to truly make unmanned aerial vehicles to be autonomous. As the starting point, in this paper, we have selected genetic algorithm as the method to achieve autonomy and will check the possibility of self-regulated autonomous unmanned aerial vehicle by applying the genetic algorithm. Landing has always been one of the most important tasks for aerial vehicles. Particularly, self-regulated autonomous landing is essential when it comes to unmanned aerial navigation. Since researching the autonomy of landing through falling body installed with a lunar Lander-like propulsion system would be more efficient for attaining the generalization of autonomous landing, it is applied on the simulation of computer-simulated falling body. When applying genetic algorithm, by first encoding genome into only 4 types of actions(left turn, right turn, thrust, free fall) then applying it on unmanned falling body, and finally combining the major computations of genetic algorithm to the unmanned falling body, we have made a successful progress in experiments. Meanwhile, previous studies have relied on various sensors to correct vertical, rotational, and horizontal errors. However, the use of measurement sensors has to be minimized in order to achieve true autonomy. The greatest accomplishment in this study was implemented a computer-simulated unmanned aerial vehicle, in order to minimize reliance of sensors, which can achieve the true meaning of autonomy of unmanned aerial vehicle and establishing a test bed for verifying the possibility by using genetic algorithm.

Keywords Unmanned aerial vehicle · Genetic algorithm · Simulation · Landing autonomy

C. Han (✉)
Department of Computer Science, Korea Military Academy, 574 Hwarang-ro, Nowon-gu, Seoul, South Korea
e-mail: chhan46@gmail.com

© Springer Nature Switzerland AG 2020
R. Lee (ed.), *Computational Science/Intelligence and Applied Informatics*,
Studies in Computational Intelligence 848,
https://doi.org/10.1007/978-3-030-25225-0_4

49

1 Introduction

The ultimate goal of this current research through this present paper is to accomplish a real autonomy of unmanned aerial vehicle (UAV). As a starting step of such object, in this paper, a genetic algorithm is selected for the methodology of the autonomy accomplishment and by applying the methodology we wants to verify the possibility of UAV autonomy. In terms of verifying a success possibility of a system, applying a computer-simulated system can be a starting point. In addition, among take-off, navigation, and landing in the UAV's various behaviors, the landing behavior is applied onto the computer-simulated system. Thus this paper is to verify the success possibility of autonomous landing, applying the genetic algorithm to the computer-simulated UAV (CS UAV) by computer simulation. As a prior step before producing an autonomous CS UAV, a man-operated dropper is generated for the CS UAV production and computer simulation to be easier. Most of the modules being produced in the production process of the manned dropper are used in the process of producing the autonomous CS UAV. In addition, an additional effect of experiencing a difficulty when a human subject operates the manually-operated dropper to be successfully landed can be obtained.

Landing is one of the most important classical tasks in the domain of aerial vehicle. As recently the development and application scope of UAV such as drone [1–4] are extended, its importance is being highlighted. It is the autonomy that is the necessary element to satisfy the scope extension in UAV's application and usage. Toward a landing spot, after a UAV navigates in some range, autonomously safe and correct landing down to the spot is one of the most important factors in UAV autonomy which is the ultimate and intrinsic objective of the unmanned aerial vehicle.

In many pervious research [1–4], a representative method for an UAV landing is to make it to navigate right above the landing pad, then to reduce its altitude, while maintaining itself over the pad. However their research is insufficient in making their landing to be a true meaning of autonomy. In order to reduce the altitude while positioning itself over the landing spot, the UAV must know how many distances remain both horizontally and vertically, and how much orientation should be modified toward the spot. They use various sensors for their landing. The values from the sensor reading are directly used to adjust the errors of horizontal distance, vertical one, and orientation. But in the 3 errors, the horizontal distance one is less dependent on the direct sensor reading. The horizontal distance error is continuously modified like in Fig. 1 by applying the pixel numbers taken by its camera sensor into a proportional equation of trigonometric function.

The study in this paper, being different from previous other researches, is to search a methodology where UAV is not dependent on sensors. Because the falling dropper in this paper is in a virtual simulation world, and not in the real one, a complete comparison is not possible. However, although a real UAV includes a complexity where various real parameters such as weather condition are considered, this current study has a meaningful contribution based on a fact that the sensor dependency should be minimized if we really want to accomplish a true meaning of autonomy.

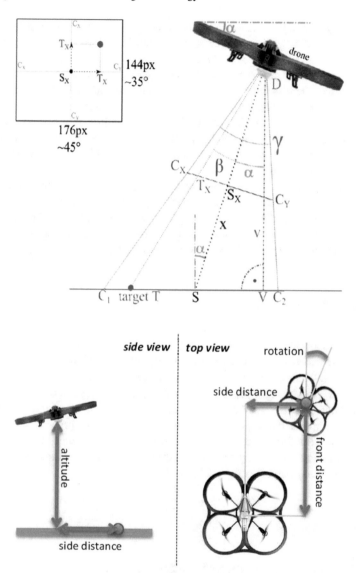

Fig. 1 Top image shows a classical landing mechanism of drone with error modification for horizontal distance [3], and bottom image shows relative position of target and drone [2]

In landing of aerial vehicles, the lunar landing [5] is one of the most historical events. The ultimate goal in this current research is to make any aerial vehicle to be autonomously landed. Because lunar landing can be a role-model in a domain of landing, it is meaningful in experiments by simulating the lunar landing mechanism. Also, the experiment is done based on position coordinates with x-axis and y-axis in 2-dimensional computer screen. Although some parameters such as the weather-related one can be considered in a real UAV experiment, the fact that only the position coordinates is considered is not a big problem because the objective of this current study is to verity the possibility where the genetic algorithm can be contributed in accomplishing autonomy.

In Sect. 2, a major process of constructing a computer-simulated UAV7 is explained. In Sect. 3, how to apply genetic algorithm is addressed. Through this application, the dropper can autonomously falls, not through man-manipulated operation. Results of experiments from man-operated dropping to autonomous CS UAV are shown in Sect. 4 and conclusion is written in Sect. 5.

2 CS UAV Construction

Because 3-dimensional real world is simulated on a 2-dimensional computer screen, the front and back in a horizontal distance is naturally omitted from all the 3-elements (i.e., horizontal distance, vertical one, and orientation one).

In this section, a major process of generating the CS UAV and its simulation is addressed and after the generation, a human subject can manually operate the CS UAV. A man-operated dropper will be constructed which is a process in the middle of completing the unmanned dropper controlled by genetic algorithm. By letting subjects to operate the falling dropper, an effect of experiencing that the dropper's landing is not easy can be extracted.

2.1 Components

Figure 2 shows an experimental environment of CS UAV.

The experiment environment is made up of a dropping base, a landing pad, a dropper, and the dropper is then composed of its body and a thrust flame. The flame is on only when a subject presses a specific keyboard or the CS UAV needs to be thrust. In the computer-simulated unmanned aerial vehicle in this study, the flame is autonomously on when thrust in the found optimal action sequence is required.

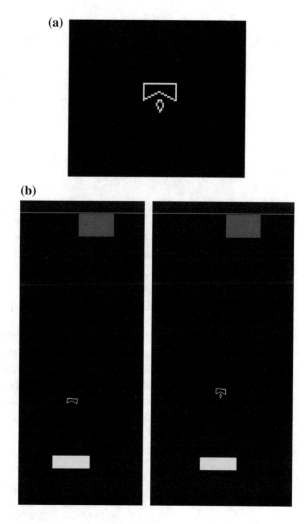

Fig. 2 Falling body experimental environment (Base, Falling body, and Landing pad). **a** represents the falling body with a thrust flame. A thrust action performed in the image on the right in (**b**)

2.2 Action

As methods of the CS UAV for modifying a horizontal distance, a vertical one, and an orientation one, left-turn, right-turn, and thrust are sufficient as actions. Because the vertical distance is modified by the free-fall of the fallen dropper, the CS UAV in this paper is also equipped with a free-fall mechanism. The oriental distance modification is naturally dealt with left-turn or right-turn. Thus the functions are embedded into left arrow and right one. The horizontal distance modification can be done by combining left-turn/right-turn with thrust.

For an object to freely behave a transformation module is required [6].

Rotate, scale, and translate are as follows.

$$[x\,y\,1]*\begin{bmatrix} \cos\theta & \sin\theta & 0 \\ -\sin\theta & \cos\theta & 0 \\ 0 & 0 & 1 \end{bmatrix} = [x*\cos\theta - y*\sin\theta \quad x*\sin\theta + y*\cos\theta \quad 1]$$

$$[x\,y\,1]*\begin{bmatrix} k & 0 & 0 \\ 0 & k & 0 \\ 0 & 0 & 1 \end{bmatrix} = [x*k \quad y*k \quad 1]$$

$$[x\,y\,1]*\begin{bmatrix} 1 & 0 & 0 \\ 0 & 1 & 0 \\ a & b & 1 \end{bmatrix} = [x+a \quad y+b \quad 1]$$

If the transformation is represented by a matrix like Fig. 3, an efficiency of computation can be improved because the 3 components are collectively dealt at once.

The distance from the dropping base to the landing zone is the length of Y-axis, which are 600. The time to reach the ground can be obtained by the equation below.

$$d = ut + at\,2/2,$$

u: initial speed, a: accelerate speed

Because the earth gravity is 9.8 m/s^2, the time to take a length of 600 pixels is 11 s, which is somewhat long to the audience. On the other hand, if this experiment is assumed to occur in the Moon, because the moon gravity is 1.63 m/s^2, the time is 27 s, which is very long to the audience. A parameter is introduced for the audience to realistically feel the time to reach the target. 1 pixel is assumed to be 1 m since the computation is to compare the simulations through the gravity difference between moon and earth.

Fig. 3 Major module for left-rotate, right-rotate, and thrust

```
if (f_space==true) {
    double ShipAcceleration =(THRUST_PER_SECOND
* TimeElapsed) / Mass;
    Velocity.x -= ShipAcceleration * Math.sin(Rotation);
    Velocity.y -= ShipAcceleration * Math.cos(Rotation);
    JetOn = true;
}
if(f_left==false && f_right==true) {
    Rotation -= ROTATION_PER_SECOND * TimeElapsed;
    if (Rotation < -PI){Rotation += 2*PI;}
}
if(f_left==true && f_right==false) {
    Rotation += ROTATION_PER_SECOND * TimeElapsed;
    if (Rotation > 2*PI){Rotation -= 2*PI;}
}
```

2.3 Test to Check if Landing Is Successful

When the dropping object reaches a designated height, a test whether the landing is successfully done is performed. The successful landing is considered when the distance of centers between the dropping object and the landing zone, and the dropping object's speed and rotation distance are within specified ranges.

After a series of simulation process and the test for successful landing are completed, if a reset module is performed the landing zone is positioned in a random location and then the dropping operation can be done again.

3 Genetic Algorithm Application

In this section, a production of an autonomous CS UAV is completed by applying genetic algorithm into the dropping object made in the previous section.

3.1 Genome Encoding

A key of this section is to decide gene characteristics in which the CS UAV's behavior during its life cycle is determined in advance. All the required actions being possessed when the CS UAV is born only are left-turn, right-turn, thrust, and free-fall. The other element related to how the CS UAV lives while drawing a trace of life is how long each action is represented. Thus the two elements, action and duration (A and D) is considered as a gene. The length of genome is experimentally made up of 20 genes (Fig. 4).

3.2 Selection of Successor Group

The population, a quantity of genome which participates in the genetic algorithm learning process is 100. The population is updated by dominant genes as each generation goes by. Roulette-wheel selection method is adopted to select the dominant genes. The size of fitness of each genome is used to be selected as a selection probability. The bigger the fitness is the more the probability to be selected is. This kind of mechanism can assure to hand over dominant genes to next generation in reasonable and fair manner (Fig. 5).

Fig. 4 Genome data structure

```
Genome RouletteWheelSelection(){
    double Slice = new Random().nextDouble() * m_dTotalFitness;
    double Total    = 0.0;

    int SelectedGenome = 0;

    for (int i=0; i<m_iPopSize; ++i){
        Total += vecGaPop.get(i).dFitness;
        if (Total > Slice) {
            SelectedGenome = i;
            break;
        }
    }
    return vecGaPop.get(SelectedGenome);
}
```

Fig. 5 Major mechanism for the selection of successor group

Fig. 6 Mutate mechanism
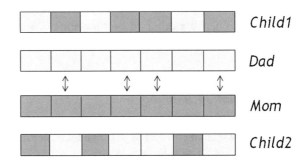

3.3   Cross-Over and Mutation

A basic operation to produce new chromosome is to cross-over the parent's ones. There are some methods in the cross-over process. Multi-point cross-over operation is used in this current research. The basic process of cross-over is to set one parent to 1st child and the other parent to 2nd child. However, if a designated cross-over rate is met at each gene, each child takes genes from the opposite side of parent.

Mutate operation has a positive effect to produce new children. In this current paper, the mutate rate is set to 0.01 and thus if each gene is in this rate, the action and the duration of the gene is changed with random values (Fig. 6).

3.4   Genome Decoding and Object Function

If all the process, the effort to the generation improvement, i.e., gene encoding, selection of new dominant genes, and cross-over and mutate is accomplished, then now it is time to make each chromosome to be performed. The left side is an image of encoded genome. Each one of 100 members in the population possesses 20 genes (Fig 7).

Fig. 7 Genome encoding
(a) and decoding (**b**)

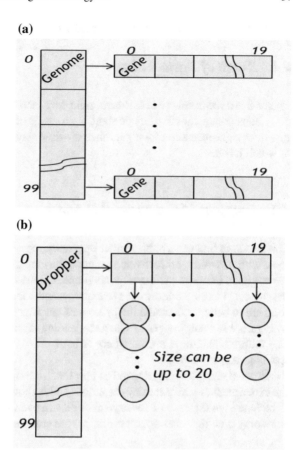

The right-side is an image of a decoded stage into a dropping object. The maximum size of action is 20, and thus the action size of each gene is constructed same as the size of its duration. For example, the 1st CS UAV begins with 3 times of left-turn, and on the other hand, 2nd CS UAV begins with 7 times of right-turn.

The object function each chromosome obtains is as follows.

$$F(d, s, r) = \{(d < 20)\ \&\&\ (s < 0.5)\ \&\&\ (r < pi/16)\}$$

Here, d is the distance to the landing pad, r means rotational distance.

4 Landing Experiment and Result

4.1 Result of Basic Action

Figure 8 is a result image of left-turn, right-turn, and thrust.

A behavior of moving left or right is an event of combination of left-turn and thrust, or right-turn and thrust, and thus the dropping object moves to the direction to which it rotates.

4.2 Result of Man-Operated Dropper

The reason to first construct the man-operated dropper and then execute the experiment with it, avoiding directly constructing the unmanned one, is that most modules are common between the man-operated one and the unmanned one and that letting the subjects including developers to experience the landing mechanism is necessary for them to better understand the process of landing.

Figure 9 is result images of when the landing is successful and is not successful. The failure situation is a case where at least one of the 3 tolerances is not in the specified value (Table 1).

Although the result may depend on how much a subject is studied through training, the experiment is performed by the author and it shows 45% success rate. The time to be landed in the case of success takes 48 s in average. Modules sufficient to add autonomy onto this man-operator dropper are completed. Through this kind of pro-

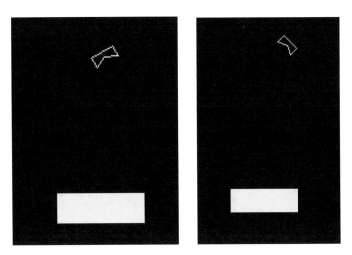

Fig. 8 Actions of left-rotate and right-rotate. The thrust can be checked in Fig. 2

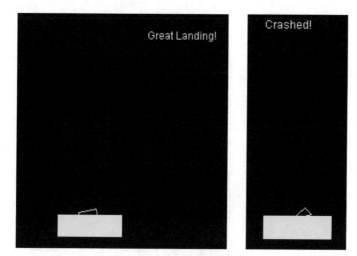

Fig. 9 Success landing on the left, and failure one on the right

cedure, an additional outcome that subjects can experience a difficulty in producing an autonomous CS UAV can be obtained.

4.3 Result of Autonomous CS UAV

The 10 experiments were conducted. The training in each experiment was done in 400 generations (1 epoch). If the landing is not successful in one epoch, the CS UAV system automatically is reset, the landing pad is randomly relocated, and the CS UAV autonomously learns the new world and is again trained by itself (Table 2).

The landings through learning-training process were successful in 7 out of 10. The number of generation required to be trained is 164 in average. Even when the landing was not successful in the one epoch, it is successful in the next epoch. The failure was the case where landing pad is created at far-left or far-right to the screen, thus the horizontal distance to the dropping base was too long. In such case, it is natural for the learning process not to be easy. When the horizontal distance to a target position is long, it will be better for an UAV to navigate further to the target before a landing mechanism occurs. If these points are considered, the learning-training mechanism in this current paper brings quite successful and robustness result to the CS UAV system.

Table 1 Result of manned landing

th	1	2	3	4	5	6	7	8	9	10
s	51	46	33	53	35	29	49	38	38	67
o/x	O	X	O	O	X	X	X	O	X	O
th	11	12	13	14	15	16	17	18	19	20
s	38	44	46	45	56	50	61	34	45	30
o/x	X	X	X	X	X	O	O	O	O	X

Table 2 Result of unmanned landing

th	1	2	3	4	5	6	7	8	9	10
Epoch	101	272	111	–	193	242	–	94	–	135
o/x	O	O	O	X	O	O	X	O	X	O

5 Conclusion

In this paper, a simulated UAV is constructed that can perform a modeling and computer-simulation of autonomous landing study of UAV. By making human subjects to operate the CS UAV, an additional outcome that subjects can experience a difficulty in producing an autonomous CS UAV can be obtained. Experiments were successful after the genome encoded with only 4 actions (left-turn, right-turn, thrust, free-fall) and major operations of genetic algorithm were applied to the CS UAV. The major contribution of this paper is that a computer-simulated UAV which can minimize a dependency to sensors and thus can achieve a true meaning of autonomy of UAV is constructed. A test-bed for checking the autonomy possibility through applying genetic algorithm has been constructed. By this, a fact that genetic algorithm can be a major methodology to achieve autonomy is verified. Thus, this paper is offering a milestone to UAVs' autonomy and makes a contribution in obtaining real unmanned-ness.

Acknowledgements The project described here has been sponsored by the Hwarangdae Research Institute in the year of 2018.

References

1. Thiang, I.N., LuMaw, H.M.T.: Vision-based object tracking algorithm with AR. Drone. Int. J. Sci. Technol. Res. **5**(Iss. 6) (2016)
2. Bartak, R., Andrej, H., David, O.: A controller for autonomous landing of AR. Drone. In; 26th Chinese Control and Decision Conference (CCDC) (2014)
3. Bart, R., Hraško, A., Obdržalek, D.: On autonomous landing of AR. Drone: hands-on experience. In: Proceedings of the 27th International Florida Artificial Intelligence Research Society Conference (2014)
4. Jin, S., Zhang, J., Shen, L., Li, T.: On-board vision autonomous landing techniques for quadrotor: a survey. In: Proceedings of the 35th Chinese Control Conference (2016)
5. Ebner, M., Levine, J., Lucas, S.M., Schaul, T., Thompson, T., Togelius, J.: Towards a Video Game Description Language, pp. 85–100. Dagstuhl Publishing, Wadern (2013)
6. Han, C., Kim, K.M., Yoo, D.H., Eum, Y.M.: Framework integration specification necessary for simulation object rendering. In: Fall Conference of Korean Entertainment Industry Association (2017)
7. Brunelli, R., Template Matching Techniques in Computer Vision: Theory and Practice. Wiley, New York (2009)
8. Cantelli, L., Mangiameli, M., Melita, C.D., Muscato, G.: UAV/UGV cooperation for surveying operations in humanitarian demining. In: 11th IEEE International Symposium on Safety Security and Rescue Robotics, 21–26 Oct, Linkoping, Sweden (2013)
9. King, M.: Process Control: A Practical Approach. Wiley, Chichester, UK (2010)
10. Lange, S., Sunderhauf, N., Protzel, P.: A vision based onboard approach for landing and position control of an autonomous multirotor UAV in GPS-denied environments. In: International Conference on Advanced Robotics, pp. 1–6 (2009)
11. Russell, S., Norvig, P.: Artificial Intelligence: A Modern Approach, 3rd edn. Prentice Hall, Englewood Cliffs, NJ (2010)

12. Serra, P., Cunha, R., Hamel, T., Cabecinhas, D., Silvestre, C.: Landing of a quadrotor on a moving target using dynamic image-based visual servo control. IEEE Trans. Robot. **32**(6), 1524–1535 (2016)

13. Szeliski, R., Computer Vision: Algorithms and Applications. Springer, Berlin (2010)

14. Yang, S., Scherer, S.A., Zell, A.: An onboard monocular vision system for autonomous takeoff, hovering and landing of a micro aerial vehicle. J. Intell. Rob. Syst. **69**(1–4), 499–515 (2013)

15. Yilmaz, A., Javed, O., Shah, M.: Object tracking: a survey. ACM Comput. Surv. (CSUR) **38**(4), 13 (2006)

16. Babenko, B., Yang, M.H., Belongie, S.: Robust object tracking with online multiple instance learning. IEEE Trans. Pattern Anal. Mach. Intell. **33**(8), 1619–1632 (2011)

17. Bao, C., et al.: Real time robust l1 tracker using accelerated proximal gradient approach. In: 2012 IEEE Conference on Computer Vision and Pattern Recognition (CVPR). IEEE, New York (2012)

18. Black, M.J., Jepson, A.D.: Eigentracking: robust matching and tracking of articulated objects using a view-based representation. Int. J. Comput. Vis. **26**(1), 63–84 (1998)

19. Kwon, J., Lee, K.M., Visual tracking decomposition. In: 2010 IEEE Conference on Computer Vision and Pattern Recognition (CVPR), IEEE, New York (2010)

Vision-Based Virtual Joystick Interface

Suwon Lee and Yong-Ho Seo

Abstract Input devices used nowadays are typically expensive, lack portability, and require a substantial amount of space. However, these limitations can be overcome by using virtual input devices. In this study, we propose a virtual joystick system, which is a type of virtual input device. Our system detects a handheld stick and computes the direction in which the user's hand moves relative to a user-defined center. The process of the proposed system can be divided into three stages: preprocessing, handheld stick detection, and direction calculation. In each stage, simple operations are performed and a color camera is used. Therefore, users do not need to purchase a high-performance computer or auxiliary devices to use the proposed system. We compare our system's performance, gauging accuracy, and speed with those produced using a real joystick. The proposed system's accuracy is competitive and has real-time speed in the laptop environment.

Keywords Virtual joystick · Natural user interface · Human-computer interaction

1 Introduction

Many IT companies around the world are entering the field of virtual reality (VR). Facebook bought Oculus, a head-mounted display (HMD) company, for 2.3 billion dollars; Samsung Electronics launched the Gear VR device, which uses a smartphone to create VR experience. Although these types of extensive investments and attention are in the developing stages, the importance of VR has already been empha-

S. Lee
Department of Computer Science and the Research Institute of Natural Science,
Gyeongsang National University, 501, Jinju-daero,
Jinju-si, Gyeongsangnam-so, South Korea
e-mail: leesuwon@gnu.ac.kr

Y.-H. Seo (✉)
Department of Intelligent Robot Engineering,
Mokwon University, 88, Doanbuk-ro, Seo-gu, Daejeon, South Korea
e-mail: yhseo@mokwon.ac.kr

© Springer Nature Switzerland AG 2020
R. Lee (ed.), *Computational Science/Intelligence and Applied Informatics*,
Studies in Computational Intelligence 848,
https://doi.org/10.1007/978-3-030-25225-0_5

sized. For example, VR can help avoid dangerous situations and actions by simulating them in a virtual world. These simulations include war, surgery, and piloting, which are circumstances that can be life threatening and expensive in the real world.

VR is a computer-simulated environment that simulates real or an imaginary world. In order to accurately simulate a situation, VR systems not only display the simulated environment but also involve user interaction. Therefore, input devices are necessary elements of any VR system. However, these devices may be difficult to use because they are usually expensive, lack portability, and require substantial space. Virtual input devices that can overcome these disadvantages can contribute to the vitalization of VR. In addition, virtual input devices can be used in augmented reality (AR), a type of human-computer interaction (HCI) that enhances our perception and helps us to see, hear, and feel our environments in new and enriching ways while providing local virtuality [1].

There is a variety of touchscreen mobile devices such as smartphones and tablets; therefore, a virtual joystick is usually implemented so as to be controlled by sliding fingers on a joystick icon on a touch screen. However, this requires an additional auxiliary device in the computer or laptop environment. Furthermore, sliding fingers does not mimic the feel of an actual joystick.

In this study, we propose a virtual joystick system, which is a type of a virtual input device. A high-performance computer or auxiliary devices are not required to use the proposed system. The system detects a handheld stick using a camera to determine the hand's relative movement direction based on a user-defined center. Our virtual joystick system addresses three issues, which not only overcome previously mentioned issues with actual input device but also replicate the experience of a real joystick. These three goals are:

i.   to accurately detect gestures used to control a real joystick,
ii. to create a widely usable, portable device, and
iii. to process actions in real time.

To achieve these goals, the proposed system detects a handheld stick, which is used like a real joystick, using only a webcam (typically included with the purchase of a computer and equipped in laptops) and performing simple operations; this results in fast processing.

Comparison experiments showed that the proposed system's accuracy is competitive and has real-time speed in the laptop environment. We believe that our study makes a significant contribution to the literature because this system can be used instead of physical input devices, which are expensive, lack portability, and require substantial space.

The rest of this paper is organized as follows. In Sect. 2, previous works on virtual joysticks are introduced. A detailed explanation of the process followed in our system for detecting a handheld stick and computing its direction is provided in Sect. 3. Additional functions that produce the impression of using a real joystick are presented in Sect. 4. The performance of our system is demonstrated in Sects. 5 and 6 offers conclusions and direction for future work.

2 Related Works

There have been many efforts to develop vision-based user interface systems. Many of these systems use the hand as an input device because the direct use of the hands is an attractive way to provide natural HCI. For instance, Rehg et al. [2] developed a non-invasive vision-based hand tracking system named DigitEyes, which uses line and point features extracted from the grayscale images of hands without the need for drawing and marking. The DigitEyes system has achieved tracking at speeds up to 10 Hz by employing a kinematic hand model. In contrast, Von Hardenberg et al. [3] defined real-time, barehanded interaction requirements derived from application scenarios and usability considerations. Based on these requirements, finger tracking and hand posture recognition algorithms were developed and evaluated. To prove the strength of the approach, they built three sample applications. Finger tracking and hand posture recognition were used to virtually paint on a wall, control a presentation with hand gestures, and move virtual items on a wall during brainstorming sessions.

Fingertips have been used to implement pointer-type input devices. Zhang et al. [4] developed a vision-based interface system called VisualPanel that uses an arbitrary rectangular panel and a fingertip as an intuitive, wireless, and mobile input device. The system can accurately and reliably track the panel and fingertip. By detecting click and drag behavior, the system enables users to perform many tasks, such as remote control of a large display or typing on a simulated keyboard. Users can naturally use their fingers or other tip pointers to perform commands and enter text. In addition, Iannizzotto et al. [5] developed a vision-based pointer device named the Graylevel VisualGlove, which can reliably and efficiently replace common pointer devices such as a mouse or stylus in most situations without requiring high computing power or expensive additional hardware. The system relies only on gray level gradients and can operate even under extreme lighting conditions. Niikura et al. [6] developed a vision-based 3D input interface for mobile devices that do not require a wide space on the device's surface, other physical devices, or specific environments. Users can operate the device by moving their fingertip in the air. Because a fingertip moves quickly around a camera, a high-frame rate camera was necessary for stable tracking. Finally, Lee et al. [7] proposed a vision based method that recognizes finger actions for use in electronic device interfaces. Fingertip tracking is implemented by detected region based tracking. The parameters of the fingertip such as position, thickness, and orientation are calculated by analyzing the contour of the tracked fingertip. Finger actions, such as movement, clicking, and pointing, are recognized by analyzing the parameters of the fingertip.

Other researchers have concentrated on mouse-like devices. A computer vision-based mouse that can control and command the cursor of a computer or a computerized system using a camera was developed by Erdem et al. [8]. To move the cursor on a computer screen, the user simply needs to move a mouse-shaped manual device placed on a surface within the viewing area of the camera. The images produced by the camera were analyzed using computer vision techniques, and the computer moves the cursor according to mouse movements. The computer vision-based mouse has areas corresponding to the click buttons. To click a button, users simply cover one of these areas with their fingers. In addition, Niyazi et al. [9] developed a mouse simulation system that controls the mouse cursor movement using a real-time camera. In the system, recognition and pose estimation are user independent and robust because the system employs color tapes on the user's finger to perform actions. The system can be used as an intuitive input interface for applications that require multidimensional control. Lastly, Banerjee et al. [10] developed a virtual mouse system in which users control the cursor movement and click events of the mouse using hand gestures that were recognized by color detection technique. Their aim was to develop a cost-effective virtual HCI device using a web camera.

Systems have been developed for specific scenarios. Graetzel et al. [11] developed a system that uses computer vision techniques to replace standard computer mouse functions with hand gestures. The system is designed to enable contactless HCI, allowing surgeons to use computers more effectively during surgery. Moreover, Robertson et al. [12] proposed a vision-based virtual mouse interface using a robotic head, visual tracking of users' head and hand positions, and recognition of user hand signs to control an intelligent kiosk. The user interface, among other things, supports the smooth control of the mouse pointer and buttons using hand signs and movements. The algorithms and architectures of the real-time vision system and robot controller were described.

Natural, fingertip-based interaction with virtual objects in AR environments has also been considered. Buchmann et al. [13] used image processing software and finger- and hand-based fiducial markers to track users' gestures, stencil buffering to enable a user to always see his/her fingers, and fingertip-based haptic feedback devices to enable users to feel virtual objects. Another AR-based user interface system for mobile devices was developed by Higuchi et al. [14] Using AR technology, the system augments virtual objects on real images captured by a camera attached to the back of a mobile device. The system allows users to operate the mobile device by manipulating the virtual objects with their hand in the space behind the mobile device.

Wilson et al. [15] developed a computer vision-based pointing device and gesture input system called FlowMouse that uses optical flow techniques to model hand movements and a capacitive touch sensor to activate and deactivate interaction. Because it uses optical flow rather than more traditional tracking-based methods, FlowMouse is very robust, simple to design, and offers opportunities for fluid gesture-based interactions that are more sophisticated than just the emulation of mouse-like pointing devices. Hand gestures were also used by Argyros et al. [16] to develop two HCI systems. Specifically, two different hand gesture vocabularies were proposed to enable the remote manipulation of a computer's mouse. One of the developed vocabularies is based on static 2D hand postures, while the second relies on 3D information and uses a mixture of static and dynamic gestures. Both interfaces were extensively tested and their relative strengths and weaknesses were evaluated. Overall, the proposed approaches are robust and can support vision-based HCI in real-world situations.

To detect the specific gesture of pinching (when a user touches their thumb and forefinger together), Wilson et al. [17] proposed a computer vision technique for close-range and relatively controlled viewing environments. The technique avoids complicated and fragile hand tracking algorithms by detecting the hole formed when the thumb and forefinger are touching; this hole is found by simple analysis of the connected components of the background segmented against the hand.

Other researchers have concentrated on simulating keyboards for user input. Habib et al. [18] developed a mono-vision virtual keyboard system for consumers of mobile and portable computing devices. They employed fuzzy approaches for gesture recognition to analyze a user's hand and finger gestures captured in a video sequence and reveal the key pressed on a printed sheet keyboard. Du et al. [19] developed a virtual keyboard system by proposing a multi-level feature matching method for reconstructing 3D hand posture. The human hand was modeled by mixing different levels of detail, from skeletal to polygonal surface representation. Different types of features were extracted and paired with the corresponding model. In addition, Murase et al. [20] developed a gesture-based virtual keyboard with QWERTY key layout that requires only one camera. The virtual keyboard tracks the user's fingers and recognizes gestures as the input, and each of its virtual keys follows a corresponding finger. Therefore, even if the hands are replaced during input, the user can input characters at his/her user preferred hand position. Finally, Yousaf et al. [21] developed a virtual keyboard using a keystroke detection and recognition approach based on finger joint tracking. A flat surface is used as the workspace for the keystroke recognition, replacing the actual keypads. The activities of user's hands are captured in a video sequence for locating and tracking the finger joints. Their movements are analyzed using computer vision algorithms to detect and recognize the keystrokes.

Systems specifically for disabled users have also been developed. Varona et al. [22] developed a vision-based user interface designed to achieve computer accessibility for disabled users with motor impairments. The interface automatically finds the user's face and tracks it over time to recognize gestures within the face area in real time. They showed how the system could be used to replace a conventional mouse device for computer interaction and as a communication system for nonverbal children. Wang et al. [23] developed a six degree-of-freedom virtual mouse based on hand gestures. The hand tracking and gesture recognition framework includes the motion, skin color, and finger information of the hand. It can even provide computer access for severely disabled users who cannot bend their fingers at all. Pugeault et al. [24] developed an interactive hand-shape recognition user interface system for American Sign Language (ASL) finger-spelling. The system employs a Microsoft Kinect sensor to detect and track the user's hand. The system works in real time and is integrated into an interactive user interface so that the signer can select between ambiguous detections. In addition, the system is integrated with an English dictionary for efficient writing.

3 Proposed Methods

In the proposed system, the process followed to determine the virtual joystick's direction using camera input can be divided into three stages. First, preprocessing is done to extract an image of the skin of the user's hand. In the second stage, a handheld stick in the skin image is detected. The last stage establishes the virtual joystick's direction by calculating the relative direction of the handheld stick based on a user-defined center. Each stage requires performing simple operations; an outline of the proposed system is illustrated in Fig. 1.

3.1 Preprocessing

In the preprocessing stage, background subtraction and extraction of skin color are applied to the camera input. This reduces the target area to be considered for post-processing.

In order to remove the background, the virtual joystick system uses a simple method. Although many background subtraction techniques can be used, they typically require multiple image frames to create a background model. In order to obtain these background images, the system needs to utilize a previously captured background or a new one. However, the proposed system should be capable of operating in a laptop environment, in which case the background frequently changes. Therefore, background captured before running the virtual joystick will most likely not be useful. On the other hand, requiring users to wait while images are obtained does not align

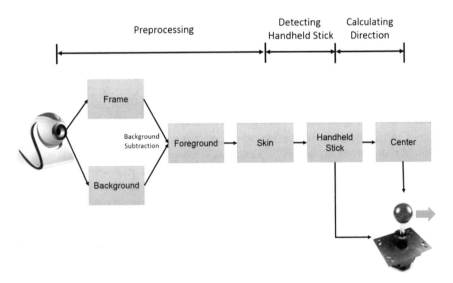

Fig. 1 Overview of the proposed system

with the system's goal of fast processing. Another motivation behind the use of a simple background subtraction method is to reduce the target area for post-processing, which is greatly diminished after the virtual joystick's center is established and useless background information is removed. Therefore, the background is determined and set in the first frame, and the foreground is defined by the difference between the background and current frame, if the difference exceeds a certain threshold. This threshold is not zero so that it is robust to noise and illumination.

The skin image extracted from the foreground is processed after converting the RGB color space to an HSV color space because HSV is more robust to illumination than RGB. Skin color is previously learned and assigned a value. However, if there is a difference between the learned color and the user's actual skin color due to which the virtual joystick system is unable to recognize the color, the user can select the color in the captured image. Automatic identification of user's skin color will be a part of future work.

3.2 Detecting the Handheld Stick

The role of the detection stage is to determine the area that meets the handheld stick's conditions using the skin color extracted in the previous stage. To find this patch, we adopt the sliding window method.

The requirements of a handheld stick are as follows. First, the center of the window should be the stick. Second, the hand should be within an eight-point perimeter of the center. Third, the color of most of the area within the window should be that of the skin. Last, the diameter of the stick should exceed a threshold value.

Calculating the stick's diameter includes checking the pixels of the diagonal that connect the upper left and lower right corners. On the diagonal, only the segment that includes the center and is not of skin color is considered as the stick's diameter. The red line in Fig. 2 illustrates this concept. The parts of the diagonal that are of skin color or do not include the center do not belong to the diameter. This is represented by the red dotted lines in Fig. 2. A visualization of all the requirements is shown in Fig. 2.

In addition to detecting the hand, this stage establishes the virtual joystick's center. After detecting the hand, the system determines whether the current hand position is close to the hand position in the previous frame. If this is true for multiple frames, the center is established to be the position in the last frame. After setting the center, the sliding window narrows the applicable area from the skin area to the area near the center. This process greatly reduces the target area use to detect the handheld stick.

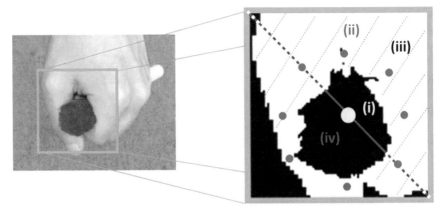

Fig. 2 Window requirements for the handheld stick are: (i) the center color should be of the stick and not skin color (yellow circle); (ii) eight arbitrarily defined points should be of skin color (blue circles); (iii) a large portion of the window should include skin color (black dotted lines); (iv) the stick's radius should exceed a threshold value (red line)

3.3 Calculating the Direction

The last stage determines the virtual joystick's direction by calculating and quantizing the angle between the current handheld stick position and the center. The angle is formed between the line segment connecting the current stick position with the virtual joystick's center and the vertical line that passes through the center. The angle between the current handheld stick position and the center is calculated as follows:

$$\theta = \tan^{-1} \frac{center_x - hs_x}{center_y - hs_y} \tag{1}$$

where *center* is the center position and *hs* is the current handheld stick position. After calculating the angle, it is quantized into eight directions. The two steps of this process are illustrated in Fig. 3.

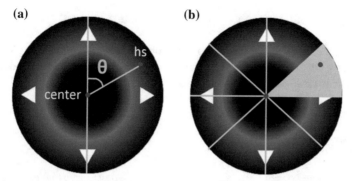

Fig. 3 a Calculation of the angle between the current handheld stick position and the center; **b** final direction determined after angle calculation

4 User Customization

When users purchase a real joystick, the program that offers the joystick's parameter-changing functions is usually enclosed as a consideration for different play styles of users. Our virtual joystick also offers these functions, which are discussed in the following subsections.

4.1 Dead Zone

If the joystick moves in a specific direction every time a user's hand moves slightly, such as when using with a trembling hand, unintentional directions can be detected. Therefore, joystick movement within a certain range from the center is ignored. This range is called the dead zone. The dead zone's range varies depending on the user and the joystick. Dead zone adjustment is an essential part of performance optimization; many studies on joystick usage have already considered this issue [6]. In the proposed system, the dead zone can be set according to the user's inputted radius.

4.2 Response Speed

In order to output the same direction multiple times, users tilt the joystick. However, if the amount of time required to output a certain direction is too small, accidental minor tilts will be incorporated. On the other hand, if this time is too long, users must wait in order to achieve the desired result. Therefore, response speed adjustment for direction output must be included in order to create a convenient virtual joystick. In the proposed system, users input the speed; the unit of time required for direction output is calculated by subtracting this speed from the input frame rate as follows:

$$\text{unit\_of\_time} = (\text{input\_frame\_rate}) - (\text{speed}). \qquad (2)$$

Users can select a speed from zero to the input frame rate minus one. Therefore, it produces a direction printing unit of time with a value between the ratio of one and the input frame rate and one second, inclusive.

5 Experiments

The purpose of our experimentation is to demonstrate that our virtual joystick system imitates a real joystick. We evaluate the accuracy and speed of the proposed system. The results suggest that our virtual joystick system has a similar accuracy to a real joystick and operates in real time.

5.1 Accuracy

In order to verify the accuracy of the virtual joystick system, we compared the output of a real joystick with the proposed system's output. A webcam captures a user playing a game with a real joystick and our virtual joystick processes the camera input and outputs the resulting directions; this allows simultaneous and accurate comparison and we can match and compare the results of the two systems. To verify the real joystick's output, we use Joystick Tester [7], a handy application which allows you to check if your joystick is working according to its parameters. The program is designed to automatically detect the controller connected to the computer in order to provide information about its status. Figure 4 provides a visual outline of this experiment.

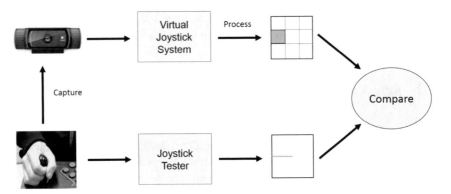

Fig. 4 Results from the virtual joystick are matched with a real joystick to ascertain the virtual system's accuracy

The resulting confusion matrix is shown in Table 1. In the table, "real" refers to the real joystick's input whereas "virtual" indicates the processing results of the camera input that captured the real joystick's control. Overall, our system has an average accuracy of 84.1% when considering all eight directions and 89.4% when considering four directions (up, down, left, and right). Results are collected during game play, which entails a high direction-change frequency. This suggests that our virtual joystick system produces results similar to that of a real joystick.

5.2 Speed

The virtual joystick system's goal is to respond to user's movements in real time. In the field of image processing and computer vision, real-time performance is considered to be one in which results are output at a rate of 20–30 fps (frames per second). This performance is easily attained using a high-performance computer. Therefore, our comparison experiment uses a low-specification laptop (1.33 GHz, Intel i3-U380UM CPU) to show that even under these conditions our system has a rate greater than 30 fps. The time required for each stage is displayed in Table 2.

6   Conclusions

This study proposed a virtual joystick system that detects a handheld stick and outputs its direction relevant to the current hand position. We focused on producing a realistic effect, equivalent to using a real joystick, to respond quickly and without the aid of additional devices. Using only color information and simple operations, we displayed the virtual joystick's direction and implemented additional functions that allow user customization.

The proposed system's performance was demonstrated using a matching method, which compared our system's results with those obtained using a real joystick and showed that our system produces results similar to that of a real joystick (with an average accuracy of 84.1%). We also assessed the real-time performance of our system.

In future work, we will add additional functions, such as pushing buttons, and implement a method that automatically defines user's skin color.

Acknowledgements This work was partly supported by the National Research Foundation of Korea (NRF) grant funded by the Korea government (MSIT) (NRF-2018R1C1B5046098) and the fund of research promotion program, Gyeongsang National University, 2018 and Mokwon University 2019.

Table 1 Multi-trial confusion matrix for virtual joystick control

Real	Virtual								
	Top left	Top	Top right	Left	Right	Bottom left	Bottom	Bottom right	Acc (%)
Top left	12	9	0	0	0	0	0	0	57.1
Top	0	94	3	0	0	0	0	0	96.9
Top right	0	6	23	0	4	0	0	0	69.7
Left	5	5	0	50	0	2	0	0	80.6
Right	0	2	10	0	96	0	0	3	86.5
Bottom left	0	0	0	3	0	12	2	0	70.6
Bottom	0	0	0	0	2	3	105	6	90.5
Bottom right	0	0	0	0	7	0	6	22	62.9

Table 2 Average processing time of the proposed virtual joystick system

	Preprocessing	Detecting the handheld stick	Calculating the direction	Total
Time (ms)	17.29	14.08	0.89	32.28
Frame rate	57.84	71.02	1123	30.98

References

1. Lee, S., Jung, J., Hong, J., Ryu, J.B., Yang, H.S.: AR paint: a fusion system of a paint tool and AR. In: International Conference on Entertainment Computing, pp. 122–129. Springer, Berlin, Heidelberg (2012)
2. Rehg, J.M., Kanade, T.: Digiteyes: Vision-based hand tracking for human-computer interaction. In: Proceedings of 1994 IEEE Workshop on Motion of Non-rigid and Articulated Objects, pp. 16–22. IEEE (1994)
3. Von Hardenberg, C., Bérard, F.: Bare-hand human-computer interaction. In: Proceedings of the 2001 Workshop on Perceptive User Interfaces, pp. 1–8. ACM (2001)
4. Zhang, Z., Wu, Y., Shan, Y., Shafer, S.: Visual panel: virtual mouse, keyboard and 3D controller with an ordinary piece of paper. In: Proceedings of the 2001 Workshop on Perceptive User Interfaces, pp. 1–8. ACM (2001)
5. Iannizzotto, G., Villari, M., Vita, L.: Hand tracking for human-computer interaction with graylevel visual glove: turning back to the simple way. In: Proceedings of the 2001 Workshop on Perceptive User Interfaces, pp. 1–7. ACM (2001)
6. Niikura, T., Hirobe, Y., Cassinelli, A., Watanabe, Y., Komuro, T., Ishikawa, M.: In-air typing interface for mobile devices with vibration feedback. In ACM SIGGRAPH 2010 Emerging Technologies, p. 15. ACM (2010)
7. Lee, D., Lee, S.: Vision-based finger action recognition by angle detection and contour analysis. ETRI J. **33**(3), 415–422 (2011)
8. Erdem, A., Erdem, E., Yardimci, Y., Atalay, V., Cetin, A.E.: Computer vision based mouse. In: 2002 IEEE International Conference on Acoustics, Speech, and Signal Processing, vol. 4, pp. IV-4178. IEEE (2002)
9. Niyazi, K., Kumar, V., Mahe, S., Vyawahare, S.: Mouse simulation using two coloured tapes. Department of Computer Science, University of Pune, India. Int. J. Inf. Sci. Tech. (IJIST) **2** (2012)
10. Banerjee, A., Ghosh, A., Bharadwaj, K., Saikia, H.: Mouse control using a web camera based on colour detection (2014). arXiv preprint arXiv:1403.4722
11. Graetzel, C., Fong, T., Grange, S., Baur, C.: A non-contact mouse for surgeon-computer interaction. Technol. Health Care **12**(3), 245–257 (2004)
12. Robertson, P., Laddaga, R., Van Kleek, M.: Virtual mouse vision based interface. In: Proceedings of the 9th International Conference on Intelligent User Interfaces, pp. 177–183. ACM (2004)
13. Buchmann, V., Violich, S., Billinghurst, M., Cockburn, A.: FingARtips: gesture based direct manipulation in augmented reality. In: Proceedings of the 2nd International Conference on Computer Graphics and Interactive Techniques in Australasia and South East Asia, pp. 212–221. ACM (2004)
14. Higuchi, M., Komuro, T.: AR typing interface for mobile devices. In: Proceedings of the 12th International Conference on Mobile and Ubiquitous Multimedia, p. 14. ACM (2013)
15. Wilson, A.D., Cutrell, E.: Flowmouse: a computer vision-based pointing and gesture input device. In: IFIP Conference on Human-Computer Interaction, pp. 565–578. Springer, Berlin, Heidelberg (2005)
16. Argyros, A.A., Lourakis, M.I.: Vision-based interpretation of hand gestures for remote control of a computer mouse. In: European Conference on Computer Vision, pp. 40–51. Springer, Berlin, Heidelberg (2006)

17. Wilson, A.D.: Robust computer vision-based detection of pinching for one and two-handed gesture input. In: Proceedings of the 19th Annual ACM Symposium on User Interface Software and Technology, pp. 255–258. ACM (2006)
18. Habib, H.A., Mufti, M.: Real time mono vision gesture based virtual keyboard system. IEEE Trans. Consum. Electron. **52**(4), 1261–1266 (2006)
19. Du, H., Charbon, E.: A virtual keyboard system based on multi-level feature matching. In: 2008 Conference on Human System Interactions, pp. 176–181. IEEE (2008)
20. Murase, T., Moteki, A., Ozawa, N., Hara, N., Nakai, T., Fujimoto, K.: Gesture keyboard requiring only one camera. In: Proceedings of the 24th Annual ACM Symposium Adjunct on User Interface Software and Technology, pp. 9–10. ACM (2011)
21. Yousaf, M.H., Habib, H.A.: Virtual keyboard: real-time finger joints tracking for keystroke detection and recognition. Arab. J. Sci. Eng. **39**(2), 923–934 (2014)
22. Varona, J., Manresa-Yee, C., Perales, F.J.: Hands-free vision-based interface for computer accessibility. J. Netw. Comput. Appl. **31**(4), 357–374 (2008)
23. Wang, X., Qin, K.: A six-degree-of-freedom virtual mouse based on hand gestures. In: 2010 International Conference on Electrical and Control Engineering, pp. 257–260. IEEE (2010)
24. Pugeault, N., Bowden, R.: Spelling it out: real-time ASL fingerspelling recognition. In: 2011 IEEE International Conference on Computer Vision Workshops (ICCV Workshops), pp. 1114–1119. IEEE (2011)
25. Ding, D., Cooper, R.A., Spaeth, D.: Optimized joystick controller. In: The 26th Annual International Conference of the IEEE Engineering in Medicine and Biology Society, vol. 2, pp. 4881–4883. IEEE (2004)
26. http://www.softpedia.com/get/System/System-Miscellaneous/Joystick-Tester.shtml

A Study on Improvement of Sound Quality of Flat Display Speaker by Improving Acoustic Radiation Characteristics

Sungtae Lee, Kwanho Park and Hyungwoo Park

Abstract With the technological improvements in the display industry, organic light-emitting diode (OLED) panel manufacturers have changed screen high-definition to simple 3D effects, augmented reality, or virtual reality. In previous research, we introduced the sound technology that vibrates the OLED panel, which name is crystal sound OLED (CSO). The difference from LCD panels, OLED panels are the simple structure, that are consists of the glass of screen and that the back-side for supports. Therefore, the screen panel can take on the role of a diaphragm of an ordinary dynamic speaker. Human hearing can confirm the sound position of the three-dimensional space by utilizing the delay difference of the sound reaching the left and right ear. If we generate these characteristics virtually, can create a 3D stereo effect. In another previous research, It is possible to create a stereo sound with multiple independent speakers. However, there is a limitation in the sound of a display speaker composed on a single panel. In this study, we analyze the acoustic characteristics of a flat speaker to realize such stereophonic sound and improve sound quality.

Keywords Direct drive speaker · Stereo excite speaker · Exciter speaker of OLED TV · 3D stereo realization

S. Lee · K. Park
LG Display, Paju-si, Gyeonggi-do, Republic of Korea
e-mail: owenlee@lgdisplay.com

K. Park
e-mail: khpark12@lgdisplay.com

H. Park (✉)
School of Information Technology, Soongsil University, 369 Snagdo-ro, Dongjak-gu, Seoul, Republic of Korea
e-mail: pphw@ssu.ac.kr

© Springer Nature Switzerland AG 2020
R. Lee (ed.), *Computational Science/Intelligence and Applied Informatics*,
Studies in Computational Intelligence 848,
https://doi.org/10.1007/978-3-030-25225-0_6

1 Introduction

People get a lot of information around them through visual, auditory, olfactory, tactile and, taste sensations. In addition, the human brain acquires information through the five senses and generates and perceives more advanced information through reasoning and imagination. In particular, audiovisual information is a very important information delivery method and has developed various techniques and methods along with human history [1]. In the past, delivering this information by itself required a high level of technology and was possible through precision parts and electrical and electronic devices. With advances in technology, the display industry has been making efforts to improve the picture quality, improve the design aspect of the screen displaying information, and improve the visual part [2]. In the process, the panel manufacturing industry has developed into the display industry, including flat display, flexible display, micro display, electronic noon, transparent display, 3D display, and mirror display [3]. In addition, the manufacturing industry that uses displays such as mobile, TV set, monitor, AR/VR, vehicle, public, advertisement screen, and a monthly touch screen is also developing [4]. Especially, the users of products and applications appearing in the downstream industries are enriched and convenient. It is also necessary for life and is a good way to display information [4].

Today, the development of wired wireless communication technology has the capacity to transmit data through a 5G communication using a super-high-resolution image taken 360°. At the beginning and end of information transmission, holograms and 3D stereoscopic video playback can be introduced, making it possible to make them look like they are in a remote space. Among them, the output of technologies such as television can be seen as a good way to store, transmit and play video and sound information. The development of technology for displaying and reproducing such information has been evolving to make the eyes and ears of a person aware of nature and reality. At the core of the technology is the ability to reproduce beyond human cognitive abilities. Utilizing the advanced technology of mankind, we can perfectly simulate a person's cognitive organs and provide more information than a sensory organ recognizes and make them feel more real than they really are. In order to realize this, it is a method of analyzing the characteristics of human cognitive organs, implementing it with more than twice accuracy and precision, transmitting information to a person, generating and processing. In this process, the elements that constitute the screen for transmitting information to the human eye have coordinated the movement of electrons in the cathode ray tube, such as a projection system using film and light, a cathode ray tube (CRT) A pixel, which is a point representing a small color, can be constituted through a liquid crystal and a color expression film such as a liquid crystal or through miniaturization and integration of a light emitting device such as a light-emitting diode. In the advanced technology, the method of composing the pixel of this information display device is the evaluation criterion of the liquid crystal type, the light emitting diode, and how to configure the light source of small size and configure the screen [1, 2].

When the image is photographed, transmitted, or stored, and reproduced appropriately, the effect of recognizing the information in the three-dimensional space can be obtained. Similarly, the sound is similarly recorded, transmitted, stored, and so on. When the sound is reproduced as a three-dimensional space according to the situation, viewers receive information in a virtual three-dimensional space. In addition to enjoying movies and dramas, it is possible to feel as if you are actually going to a place you have not visited before, and you can reproduce the performance of a famous singer who lived in another era as if you were a live person. Therefore, the display device technology can integrate the pixels of the screen to naturally stimulate a human visual system, and in the process, the screen of the information display device changes to a form in which only the screen is seen on the front side. In the past, due to the limitations of technology, it was possible to transmit low-bandwidth sounds in the past, but nowadays it has become possible to simultaneously transmit high-resolution super-high-quality pictures and more than 22 sound signals simultaneously. What is important here is to re-examine the perception of the reproduction of the screen and sound. That is, if the position of the sound that should have been generated in the image differs from the position of the sound produced by the actual sound system, the human brain accepts it as two confusing pieces of information. It is disadvantageous for accurate information transmission. In this respect, matching the position of the screen with the position of the sound is an important aspect in terms of improving the efficiency of information transmission [3–5].

Companies producing information display devices have been studying how to construct a screen considering the aesthetic aspects of the screen. In particular, how to integrate LCD and LED in a unit area to achieve high image quality is a standard that boasts a company's technical scale. However, such image-oriented technology development is somewhat neglected in the technology of sound transmission, and gradually reduces the number and location of loudspeakers, I feel the space of sound. In other words, since the key information of the three-dimensional space that a person feels is an image, the less important sound is delayed or supported from the rear. Along with this strategy, technology has changed according to the position of people who demand super-high-definition and super-large screen. Today, however, the development of reputation information display technology allows even sound to be reproduced directly on the screen. As if the sound was played directly on the screen, the technique was advanced [6–9].

Previous studies have shown that the OLED panel can be vibrated directly to reproduce sound and match the focus of the screen to the focus of the sound. We also analyzed human characteristics from a cognitive point of view. In this study, we introduce a study to improve sound quality by analyzing the shape of exciter speaker. The exciter loudspeakers, which directly oscillate the OLED screen developed in the previous research, are fixed to the back panel to fix the screen. Then, the sound is reproduced by generating vibration between the back plate and the screen in accordance with the audio signal. At this time, the OLED panel plays a role of the conduction of a general dynamic speaker, and a sound is generated. The sound will be generated, but sound quality will vary depending on the shape of the exciter and how to control it. In this study, the sound quality is improved by analyzing the vibration shape of the exciter, its own shape, and acoustic emission characteristics when several exciters are driven.

2 Previous Research and Principal Background

The sound is generated by phenomena such as friction, collision, bouncing, air flow, and explosion between an object and an object. In the process of generating sound, kinetic energy, position energy, or chemical energy is converted into another energy form, and in the process, the surrounding air molecules are vibrated together. The fundamental principle is that the vibrations of the generated air molecules propagate in the form of the longitudinal waves by pushing and pulling the surrounding air molecules and forming the wave front of the sound. And the vibrations of the air of this sect are conveyed by the medium, and it is said that the sound of the physical phenomenon heard in the ear is heard. At this time, the sound that stimulates the ear of the human being is heard as a different sound according to the generation method of the cause, the resonance obtained in the transmission process, and the characteristics of the manure, and is combined and converted by a very wide variety of methods. It will stimulate the organ. In particular, in the case of musical instruments, the types of sound generation are classified based on the excitation source caused by skipping, hitting, friction, and air flow, and the main resonance causes are used to distinguish and characterize the characteristics [10, 11].

The three elements of sound analysis are amplitude, frequency, and duration. These three elements are independent of each other, and the three elements of sound can be used to distinguish the sound of the world and to divide the characteristics. Among the three elements of sound, the size expresses the intensity of input energy into the instrument or the sounding system in the early or midst of the occurrence and determines how much energy is energized in music. The unit of measurement in decibels (dB) is used for the measurement. The human auditory characteristic has the property that the amount of change of the energy to the logarithm ratio is recognized as the isoquant, and the conversion value is obtained through the logarithmic function. The frequency is the number of times the air molecule vibrates for one second per unit time, and the characteristics between the components of the sound analyzed by frequency are independent. The generated sound is not composed of a single frequency, but the reproduction of the harmonics according to the resonance structure of the musical scale and the sound producing device according to the component corresponding to the height of the desired sound and the sounding characteristics. Frequency analysis allows the analysis of the fundamental resonant structure size, major component length, plate width, and mode of vibration [12–16].

A magnetic field is formed around a conductor through which a current flows, and an electric signal and an acoustic signal are mutually convertible according to the principle that a force is generated in a wire in a magnetic field. Of course, the energy of each has a difference, but when the signals are classified in the form of size, frequency, and duration, mutual characteristics can be preserved and converted. Particularly, a

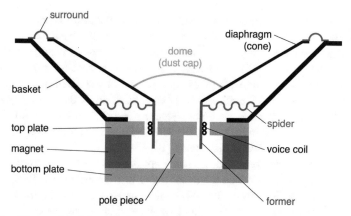

Fig. 1 Simplified model of dynamic loudspeaker [5, 18]

device for converting an electric signal into an acoustic signal is called a speaker unit. Since 1930, this unit has been developed as a basic element to regenerate sound, and to develop materials, manufacturing methods, driving methods and performance of magnetic materials [8]. In a typical, basic dynamic speaker, the voice coil moves in a magnetic field formed by a permanent magnet, which causes the cone to move back and forth, pushing and pulling air to produce a sound wave. Figure 1 shows the dynamic speaker [8]. The structure of Fig. 1 can be modeled as a 1 degree-of-freedom vibration system in which exciter, back cover, and flat plate are connected in harmony [7]. As explained above, between the magnets, the voice coils make a one-dimensional vibration, and the vibration causes the motion of the connected diaphragm. This vibration occurs inside the basket, pushing and pulling air in the direction of the front of the speaker, i.e., the dust cap. This generated vibration makes the sound. A good speaker means using less energy to make the same electrical signal as the original electrical signal. It is also said that a person may be able to implement a specific sound well by driving in a deliberate manner. In the case of a dynamic speaker, the diaphragm between the basket and the top plate is well connected to the surround and the spider, and the linear motion is well done. It operates properly in a wide frequency band without abnormal operation [5, 10, 17].

Generally, when a person is listening to music and then lowering the volume, the human ear experiences a change beyond the size of the sound. The playback sound with a low sound pressure level is low Freq. Depending on the loudness contour which indicates the human auditory characteristic. BW and High Freq. BW. The range is relatively small compared to the middle range. Therefore, the frequency characteristics of the audio system are measured and checked to see if it meets human auditory characteristics. The human ears are arranged on the left and right sides of the head. When a sound is an incident on the right side, the right ear quickly reaches a sound in time, and a difference in the level of a sound incident on the two ears due to the influence of the head shape also occurs. The left and right directions are determined by the binaural time delay and the binaural level difference. And it recognizes the sound in three-dimensional space. In the Audio System, though it gives the direction of the sound source by using it, it may give a special listening feeling by generating the unintended directional feeling due to the difference of the position and the volume of the speaker. Therefore, it is necessary to measure the directionality to judge whether it gives a proper directional feeling and to make it have a correct directional feeling and a sense of position. If the reverberation time is too much, the clarity becomes worse when the reverberation is too much, and if the reverberation is small in the music, the series of notes disappear discontinuously, and the saturation of the sound disappears [14, 17–20].

Typical dynamic speakers and flat panel speakers are different from what they are. Typical loudspeakers attach a conical paper to the vibrator to efficiently push and pull air. Flat speakers instead of this conical paper feature a flat front plate that pushes and pulls wide air to make a sound. So, instead of a cone, the voice coil motor uses a flat plate on the front to push and pull the air. Some of the voice coil motors have a flat plate and the whole unit is flat, and some voice coils are used only for piston motion. One type of flat-panel loudspeaker is a front-mounted condenser-type speaker [16, 10]. Speakers using voice coils are common and widely used, but shape and volume are disadvantageous for making a high-quality sound in a small volume [16]. In addition, in the case of a flat panel speaker, if the motion characteristics, size, and mass of the moving plate cannot be precisely controlled, it is impossible to reproduce the correct sound corresponding to the producer's intention, to reproduce the distorted sound or to reproduce the normal sound none. In this study, a vibration unit of about 1 inch is used to efficiently vibrate the LED glass layer on the front side, and the sound is reproduced by adhering between the glass plate and the backside plate holding the glass layer [11, 18, 20]. Figure 2 shows a cross-sectional model with a vibrating speaker attached between the top OLED panel and the back cover. Unlike dynamic speakers, the speaker does not have a cone shaped like a diaphragm, and a thin glass layer plays a role. In this case, the back cover and the OLED panel oscillate with each other, and the vibration occurs in the OLED with a small mass, and the sound is reproduced [18, 20, 21]. Figure 2 is a model of the direct exciter speaker of the proposed OLED display device. Figure 2 shows a single plane OLED layer touching the vibrator of the exciter. And permanent magnets and voice coils to create a linear motion of the pole-piece in the bode. The exciter is attached to the inner plate and the back-cover to absorb vibration and oscillate only the OLED layer.

As shown in Fig. 2, even if the OLED panel vibrates directly, the sound quality varies greatly depending on the shake characteristics and vibration mode character-

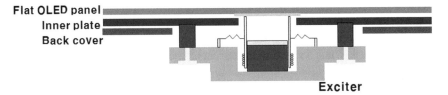

Fig. 2 A model of flat panel speaker

istics of the display. As you saw on the two flat speakers, the sound depends on the shape and characteristics of the diaphragm. In order to compensate for the disadvantages of conventional flat speakers, it is necessary to improve the sound quality by transmitting vibration to the back plate that fixes the display [15, 21]. Techniques have been applied to prevent vibrations from the left and right to realize the sound of the left and right stereo channels moving with a single plate. With this method, although one glass layer is used as a diaphragm, a flat speaker in which stereo sound is reproduced can be constituted. This study was carried out in the previous research, and the improvement of the sound quality according to the propagation of the exciter was also carried out. In this study, we evaluated the effect of encoding by adjusting the acoustic impedance to the back plate as well as the propagation model and evaluated the information transmission according to the screen configuration and the acoustic characteristic.

3   Propose Sound Enhancement

In the proposed method, we analyze the acoustic emission characteristics of the exciter as shown in Fig. 2 to realize the optimal screen and sound of the OLED. In this study, acoustic radiation patterns of a single exciter are measured and two or more exciters interact. Then, by increasing the number of exciters, the sound field is measured and the result is predicted to form an optimum sound. The acoustic measurement was carried out by measuring excitation loudspeakers, measuring the sound at regular intervals at the center point, and measuring the planar sound field even when there were two or four models. The sound source uses a white sound to confirm the response in a wide frequency range and to confirm the advantages of the arrangement and configuration. Should be treated as a 3rd level heading and should not be assigned a number.

4   Experiment and Result

The in experiment used two exciter model, one is a 10 W exciter model with two pole-pieces, and the other is a 12 W exciter model with eclipse pole-piece, to simulate acoustic radiation patterns for experiments and simulations. One was driven to confirm the radiation pattern of the proximity section. Then, two exciters were driven and their shape was measured, and the sound pressure distribution was simulated at three exciters. Experiments were carried out in a space of 30 dB (A), where the output was connected to a 20 W amplifier per each channel to generate 98 dB at the point of origin.

Figure 3 shows the sound pressure distribution of a single exciter with two pole-pieces. In this exciter model, two pole-pieces that transmit vibration to the diaphragm exhibit high sound pressure in a side-by-side direction. That is, it means that the form

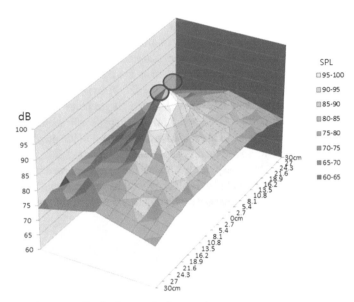

Fig. 3 A sound pressure distribution graph of one exciter

and shape of the sound source radiated according to the shape of the pole piece are different. And, in Fig. 3, a directivity angle is formed forward in a range of about 33°. This means the emission angle in the direction of the short axis when there are the long axis and the short axis of the exciter. When the radiation angle is about 33°, the shape of the sound source generated from the exciter is predicted. Figure 4 shows the result of measuring two exciters at intervals of 30 cm. It is also possible to predict that a flat sound source can be constructed by the action of two exciters as the distance between the two exciters is small. In other words, as shown in Fig. 5, it can be predicted that two or more exciters are overlapped and spread.

In the case of a model with one pole-piece in the shape of an ellipse, it is similar to the two, but the radiation angle is slightly extended. Even if the same signal is utilized, the two vibrations cause mutual interference and the radiation angle is reduced. This improvement allows sound to be supplied evenly over a wide range, and the sound field formed by four or more exciters can be seen in Fig. 6. As a result of measurement at the same height as the exciter, it is confirmed that the overall flatness is higher at 4CH than the 2CH speaker. Also, it is judged that it is possible to synthesize the sound field for four speaker sections using the system. In the case of 2CH, since there is no sound source at the center part, the effect of control decreases, and there is a limit to the directionality setting. 4CH can synthesize the position of the sound source by controlling the output and the generation time of the speaker. If the distribution of the sound source at the same height as the exciter is the same as the right picture, the sound source can be maintained by maintaining the sound pressure even at a long distance, and a wide space can be used at a sound pressure suitable for viewing the screen through plane wave reproduction.

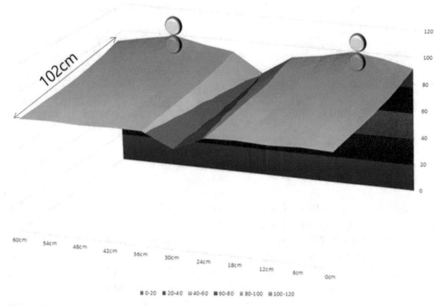

Fig. 4 A sound pressure distribution graph of two exciters

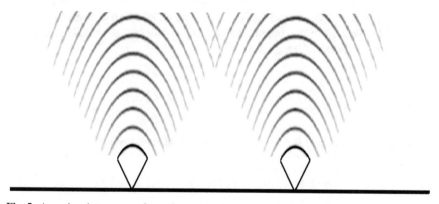

Fig. 5 An estimation pattern of sound pressure distribution with two exciters

5 Conclusion

The development of display technology and semiconductor technology has made the display system clearer and smarter. Self-luminous OLED technology also improves picture quality and improves the sound with panel exciter speakers. Previous studies have shown some good points when creating a sound by vibrating the screen. In this study, we studied the characteristics of this stimulator and studied how to make 3D stereo sound using the OLED panel and stimulator. In the experiments and results, we

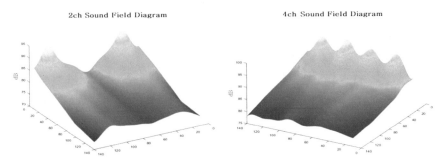

Fig. 6 Sound field measurement results of four exciter models

analyze the source emission characteristics of one exciter, analyze the characteristics of two stimuli, and increase the exciter to 4CH. The sound field generated by the stimulator and the OLED panel was measured and analyzed. We also investigated the enhancement of the sound quality of high-definition display devices using a multi-channel exciter. In addition, we have confirmed the possibility of implementing high-quality stereo sound, such as sound field synthesis. In addition, by driving four independent channels, we were able to precisely control the position of the sound and protect the screen from shrinking the sound. In the future, we will study how to improve the sound quality by actively using mutual interference of 4CH exciter speaker. Also, it is necessary to study the rendering algorithm that synthesizes the sound field so that the multi-channel output can be produced as needed according to the input sound source of the speaker.

References

1. Republic of Korea OLED Yesterday, Today and Tomorrow, Korea Display Industry Association, Sainovation (2016)
2. AMOLED Industry Market Tracker. IHS Report, 2017–2018
3. Park, Y.H., et. al.: SID 2017 (Display Week) Display industry trend vol. 17–6. PD Issue Report, Keit (2017)
4. Kim, J.T., Kim, J.H., Kim, J.O., Min, J.K.: Acoustic characteristics of a loudspeaker obtained by vibroacoustic analysis. Trans. Korean Soc. Mech. Eng. **21**(10), 1742–1756 (1997)
5. Choi, H.W., Kim, Y.J., Park, Y.W.: Acoustic analysis and vibration modeling for design of flat vibration speaker. In: Proceedings of Korean Society for Precision Engineering, 2008 Fall Conference, pp. 545–546 (2008)
6. Choi, D.J., Park, Y.W., Park, H.J.: Research of relation between sound pressure and magnetostrictive speaker for 2ch flat display. In: Proceedings of Korean Society for Precision Engineering, 2010 Fall Conference, pp. 603–604 (2010)
7. Beon, H.J.: Trends in speaker manufacturing technology at domestic and abroad. Mag. IEEE Korea **13**(6), 513–520 (1986)
8. Nam, S.H., Choi, D.J., Chae, J.S.: Objective audio quality assessment. In: Proceedings of EEE Korea, pp. 108–111 (1995)
9. Kim, J.H., Kim, J.W.: Thin speakers and transparent speakers. CERAMIST **17**(2), 60–66 (2014)

10. Lee, S.T., Park, H.W., Bae, M.J., Park, K.H.: Sound quality improvement of a flat-panel display front speaker (Emotional and EEG test of exciter speaker). SID Symp. Dig. Techn. Papers **48**(1), 1680–1681 (2017)

11. Kim, Y.H., Nam, K.U.: Lecture Notes on Acoustics. Chungmungak Press (2005)

12. Oh, S.J.: Theory and Design of Loudspeaker. SuckHakDnag Press (2011)

13. Park, U., Lee, S.B., Lee, S.H.: General Theory of Sound Technology. Cha Song Publisher (2009)

14. Park, H.W., Bae, M.J.: A study on the improvement of sound quality according to the location of OLED flat plate speaker. Asia-Pac. J. Multimedia Serv. Convergent Art Humanit. Sociol. **7**(12), 775–783 (2017)

15. Lee, S., Park, H., Bae, M., Park, K.: Study on multi-channel speakers for controlling sound field of flat panel display. J. Acoust. Soc. Am. **143**(3) (2018)

16. Park, H.W., Bae, M.J.: A study on the watching attention according to the sound position of flat panel TV. Asia-Pac. J. Multimedia Serv. Convergent Art Humanit. Sociol. **7**(7), 839–846 (2017)

17. Park, H., Lee, S., Park, K., Bae, M.: Comparison of acoustic quality according to the types of flat panel display (Direct drive with exciter speaker). Acoust. Soc. Am. **144**(3) (2018)

18. Kown, O.G.: Design of multi-channel speaker system using digital audio technology. Han-kukhaksuljungbo (2008)

19. Park, H., Bae, M., Lee, S., Park, K.: A study on the sound generation at digital signage using OLED panel. J. Acoust. Soc. Am. **143**(3) (2018)

20. Lee, S., Park, H., Park, K., Bae, M.: A study on the optimal speaker position for improving sound quality of flat panel display. Acoust. Soc. Am. **144**(3) (2018)

21. Lee, S.: Basic Properties of Sound and Application. Chung-Moon-Gak Publisher (2004)

Edge Detection in Roof Images Using Transfer Learning in CNN

Aneeqa Ahmed, Yung-Cheol Byun and Sang Yong Byun

Abstract Edge Detection in image processing is very important due to large number of applications it offers in variety of fields that extend from medical imaging to text and object detection, security, mapping of roads, real time traffic management, image inpainting, video surveillance and many more. Traditional methods for edge detection mostly rely on gradient filter based algorithms which usually require excessive pre-processing of the images for noise reduction and post-processing of the generated results in order to get fine edges. Moreover, traditional algorithms are not reliable generally because; as the noise in images increases their efficiency is affected largely due to escalation of mask size which also makes the system computationally expensive. In this paper, we will employ transfer learning in CNN method to detect edges of roof images. Incorporating CNN into edge detection problem makes the whole system simple, fast, and reliable. Moreover, with no more extra training requirement and without any additional feature extraction, CNN can process input images of any size. This technique employs feature map of the image using Visual Geometry Group (VGG) CNN network followed by application of Roberts, Prewitt, Scharr and Sobel edge operators separately to compute required edges. Interpretations of ground truths were obtained using manual techniques on roof images for performance comparison, and PNSR value of computed results via multiple operators against the ground truths is calculated.

Keywords Edge detection · Deep learning · Convolution neural networks · Roof images

A. Ahmed (✉) · Y.-C. Byun · S. Y. Byun
Department of Computer Engineering, Jeju National University, Jeju-si, Republic of Korea
e-mail: aneeqaahmed339@gmail.com

Y.-C. Byun
e-mail: ycb@jejunu.ac.kr

S. Y. Byun
e-mail: byunsy@jejunu.ac.kr

© Springer Nature Switzerland AG 2020
R. Lee (ed.), *Computational Science/Intelligence and Applied Informatics*,
Studies in Computational Intelligence 848,
https://doi.org/10.1007/978-3-030-25225-0_7

1 Introduction

Edges in images can be defined as a boundary between the object and the background which can be located by detecting the discontinuities in image brightness known as edge detection problem in image processing [1]. Edge detection is considered as one of the important phenomenon for extracting useful information from images [2] and it is broadly applied in computer vision and image processing domains. El-Sayed et al. [3] stated that crucial information in a scene is mostly hidden in its edges, thus edge detection has been a subject of vast interest to the scientists and there have been widespread research efforts in developing a good and reliable edge detection algorithm from many years. However, edges in images can be formed in variety of ways depending upon object's geometry and they may vary in shapes and sizes [3, 4]. Also, efficiency of edge detection algorithms largely depends upon the illumination conditions and noise level of the images. To devise a standard edge detection algorithm that is universally accepted is not an easy task. Traditional approaches for edge detection include gradient or difference based filters, non-maximum suppression or detection of zero crossings by Laplacian of Gaussian etc. [2, 5–7].

Traditional methods for edge detection are complex and computationally expensive as they mainly depend on computing several image features [1]. Moreover, with the increase of noise level or change of illumination conditions of images the efficiency of traditional algorithms degrades most of the times which make them highly unreliable [5]. Furthermore, many type of filters that form the basis for traditional edge detection algorithms can only detect edges in particular directions, hence it demands use of multiple filters together to detect edges in every direction. These limitations of traditional algorithms motivated researchers to incorporate machine learning techniques into edge detection problem.

In this paper, we will make use of convolutional layers of VGG CNN network [8] for computing the image feature map. We will employ transfer learning from CNN to solve edge detection problem which is a continuation of our previous work [9]. Basically, CNN is very simple and fast as it does not require any extra feature processing to detect the edges [1]. Moreover, with CNN the task of deciding and extracting the concerned features is not to be programmed manually rather CNN itself does the job and it is more efficient than traditional manual approaches. Also, unlike traditional methods CNN is strong enough to learn directly from the images and due to these capabilities, CNN can process any input image irrespective of its size and without the need of any extra training aswell [1]. Besides, it is very simple to integrate CNN into other computer vision and image processing systems so by incorporating CNN into edge detection problem, people can easily gel in this network to other applications that may require edge detection as a first step for some disease diagnosis or for image segmentation etc. [1].

It is worth mentioning that, neural networks have been exploited in many areas of image processing. In the recent past, multiple research works based on deep learning models for edge detection have been published. Some worth mentioning approaches based on CNN models for edge detection include HED, Deep Edge, Deep

Contour and Richer Convolutional Features for Edge Detection which are presented in [10–13]. However, in all of the above mentioned approaches the presence of ground truth is a mandatory requirement in order to train a model which would be able to detect the required edges. For efficient training of a model, the number of ground truths should be available in vast figures otherwise it would largely effect the accuracy of the results. Conversely, in many real time image based applications where researchers want to implement edge detection phenomenon, the availability of ground truth images mostly lacks. Even if the ground truths for a particular image data are available, the number of such images are too low for training a reasonable model. This ultimately limits the implementation deep learning based approaches for edge detection. This paper aims to propose a solution for a scenario to detect the edges using CNN model where datasets have no ground truth images available. It makes use of transfer learning method using a pre-trained CNN model. It computes the features of image using the pre-trained CNN model and further uses those strong computed features for edge detection

1.1 VGG Network

Deep Neural Networks (DNNs) have emerged as a noticeable approach in the field of computer vision and intelligent computing [14]. Researchers and engineers are motivated to design more powerful and accurate algorithms because of their increasing utilities and emerging applicability in variety of tasks. It is the architecture of neural networks that determines the accuracy of the algorithms as well it defines resource utilization. Therefore, the success of these algorithms mainly depends on the careful designing of neural network architecture.

VGG network as shown in Fig. 1 is a deep learning (DL) model which was launched in 2014 by Simonyan and Zisserman in 2014 ImageNet Large Scale Visual Recognition Challenge (ILSVRC) [8]. The architecture of VGG network depicted that network depth is very crucial and important parameter in order to achieve better performance and accuracy. VGG network is well-known for its simple architecture in view point of depth which increases progressively by adding new convolution layers using merely 3 × 3 convolution filters in all layers. This network comprises of 16 convolution layers as shown in Fig. 1 and has been trained on 4 GPUs for 2–3 weeks. VGG network has achieved remarkable accuracy in 2014 ILSVRC and team VGG network has secured top two positions in the localization and classification tasks. Along with that, VGG network is fundamentally applicable to other image processing and computer vision applications as a standard feature extractor. Weight configurations of VGG network are publically available and presently researchers are widely using this network to extract feature from images with a greater accuracy as compared to previous DL models.

The rest of the paper is organized as follows: Sect. 2 describes the proposed system for edge detection. Section 3 presents experimental results and gives performance comparison. Section 4 summarizes conclusion with a summary of future goals.

Fig. 1 VGG network
architecture

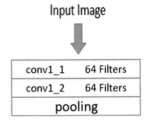

Input Image

conv1_1	64 Filters
conv1_2	64 Filters
pooling	

conv2_1	128 Filters
conv2_2	128 Filters
pooling	

conv3_1	256 Filters
conv3_2	256 Filters
conv3_3	256 Filters
pooling	

conv4_1	512 Filters
Conv4_2	512 Filters
Conv4_3	512 Filters
pooling	

conv5_1	512 Filters
Conv5_2	512 Filters
Conv5_3	512 Filters
pooling	

3 Fully Connected Layers
FC6, FC7,FC8

Output Image

2 Edge Detection Method

In this section, we have briefly discussed about the dataset that we have used in this paper, which is followed by detailed explanation of the technique being employed in this paper. Broadly, the methodology comprises of four important steps to detect the edges which are feature map computation via convolution neural network, gradient map computation, averaging all filters gradient maps, averaging all layers gradient maps.

2.1 Dataset

In this paper, we have used dataset of roof images as seen from satellite for the purpose of edge detection. The dataset used is downloaded from a renowned web repository of datasets [15] and is publically available. It contains 42,760 images of roof tops that include various types and shapes of roofs like cylindrical, square, rectangle, pointed etc. The dataset also comprises of roof images of varying illumination conditions and resolutions. Using this dataset allowed us to test our system with roof top images of variable illumination conditions with different kinds of roof shapes which increased system robustness.

2.2 Methodology

The available dataset only composed of roof images but it lacked the ground truths for edge detection. We have employed manual techniques to generate the ground truths from our datasets for few images only and these ground truths have been used during performance analysis of this method. However, to generate manual ground truths for all 42760 images is a tedious task. Ground truths are mandatory requirement if we specifically want to train a customized CNN for direct edge detection. Keeping this in view, we have incorporated an indirect method that is CNN transfer learning methodology to achieve our goal. Transfer learning allows us to use a model that is trained for one task to be reused for another allied task. We have made use of a pre-trained strong convolutional neural network namely VGG network and computed all the required features automatically from VGG network. Later on, this automatically computed strong feature map is being fed to Roberts, Prewitt, Scharr and Sobel edge detectors separately and gradients maps for every filter in each layer of VGG network is computed which are ultimately averaged out to give the total gradient maps for every layer of the network. The overall methodology is being explained in the following steps:

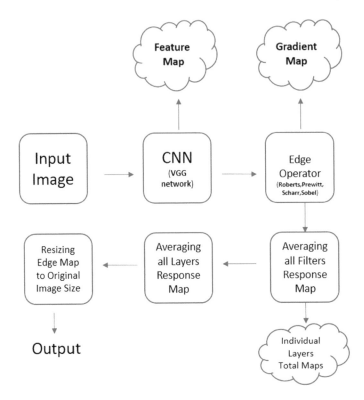

Fig. 2 Block diagram of edge detection system

(a) At the first step feature map of the input image is computed. To determine the required feature map, this approach employs the convolution layers of VGG network.

(b) Subsequently, Roberts, Prewitt, Scharr and Sobel operators are applied separately to compute gradient map for each filter in a given convolution layer.

(c) Then the average of gradient maps of all filters in a given layer is computed which gives the total gradient map for the specific layer.

(d) At this point, overall average gradient map is computed for all layers of network in order to get the finalized edge map of the image. Finally, the edge map is being resized to original input image size to visualize the resultant edges.

Figure 2 shows the block diagram of the system.

In order to investigate the response and of each layer of the network to the final outcome we have made few observations. After comparing the "step d-all layers averaged result" to the individual gradient maps of every layer in "step c" we have observed the final result for every type of image is quite degraded as compared to the manual ground truths. The reason of this type of outcome is because edge information is mainly extracted in lower layers and higher layers mainly focus towards learning

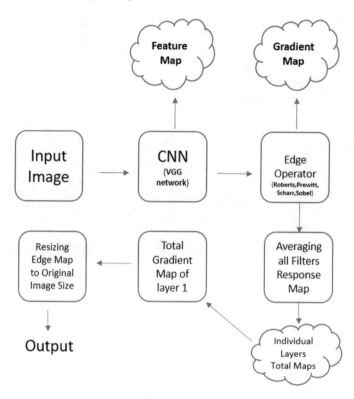

Fig. 3 Upgraded block diagram of edge detection system

of high level features like recognizing the shapes etc. The individual total gradient maps for the specific layer in "step c" have shown very promising results for edge detection in the lower layers of the network, while results gradually degrade as we move towards the higher layers of the network.

All these investigations of individual response of each layer in this approach lead us to slightly modify our methodology. Instead of computing "overall average gradient map for all layers in step d", we just picked total gradient map of all filters in layer 1 of VGG network as our final outcome for every type of image and then we resized this outcome to original input image size to visualize the resultant edges. Figure 3 shows the upgraded methodology that we followed in this work.

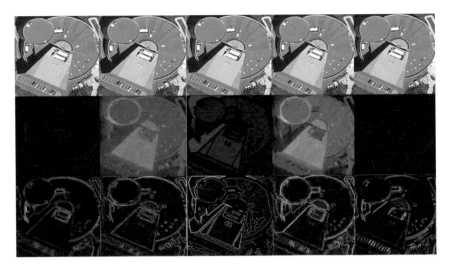

Fig. 4 Gradient map at filter level for different layers using Sobel Operator

3 Experimental Results

We have evaluated our CNN based edge detection system on the dataset of roof images provided by kaggle [15]. The results of different stages corresponding to Fig. 3 are being shown in this section. At the first step, the image is being input to the system and feature map is computed using VGG CNN network.

After the computation of feature map, Roberts, Prewitt, Scharr and Sobel operators are being employed separately to determine the gradient map of every filter in each layer of VGG network. Figure 4 displays resultant outcomes of edges for only few filters in different layers of the network with Sobel edge detector due to limitation of space. The first image in every column is the original image, followed by grey scale outcome and the last one is the resultant image in binary level.

As there are many filters in each layer of VGG network as shown already in Fig. 1, so in order to get the total response map of an entire layer, gradient maps of all filters within a layer are averaged. Figure 5 shows the complete edge maps for different layers.

From Fig. 5 it can be observed easily that gradient maps of high level layers are much degraded as we discussed earlier in Sect. 2 while discussing the individual response of each layer, so when we took the average of gradient maps for all the layers to generate a single outcome we got very degraded results Fig. 6a shows the overall Averaged Gradient Map of all layers with Sobel Operator and Fig. 5b shows the overall Averaged Gradient Map of only first 4 layers with Sobel operator. From Fig. 6a, b and from Fig. 5 we have judged that if we only consider the lowest layer that is layer 1 outcome in the final results of edge detection, we can improve our results.

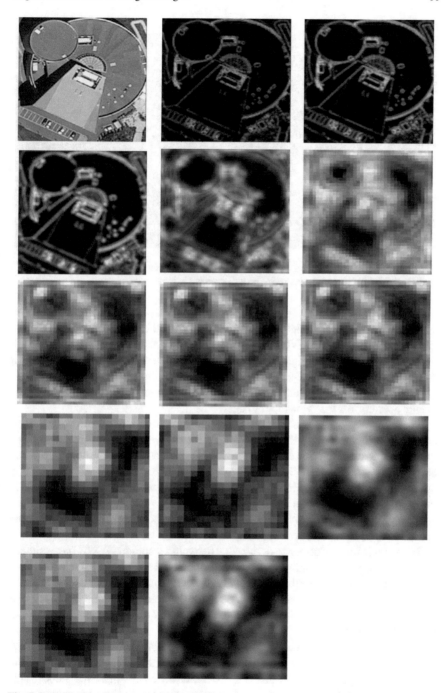

Fig. 5 Individual gradient maps for all layers using Sobel Operator

(a)   **(b)**

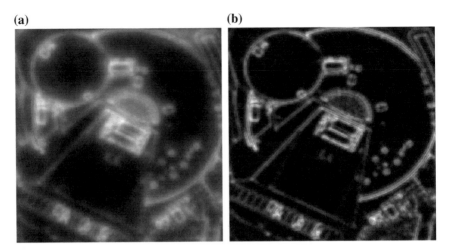

Fig. 6 **a** Overall averaged gradient map of all layers with Sobel Operator, **b** overall averaged gradient map of only first 4 layers with Sobel Operator

So, following the upgraded methodology as depicted in Fig. 3 we picked the total gradient map of layer 1 of VGG network as our resultant outcome for edges which is finally resized to the original size of the input image. Figure 7 shows the final outcome of our edge detection system with Sobel operator.

Similarly, Fig. 8 shows output of edge detection system with Roberts, Prewitt and Scharr operator for the same input image.

In Fig. 9 we have shown total gradient maps of first three layers with Prewitt and Sobel operators using this methodology for different type of roofs, also these images are of variable sizes and different illumination conditions. From Fig. 9 it can be observed that this technique performs equally good for images of any size and even with bad resolution, which is one of the major advantages of introducing CNN into edge detection problem over traditional methods.

Likewise, Fig. 10 shows final outcomes with Prewitt, Roberts, Scharr and Sobel operators for different size of roof images.

To evaluate the performance of the edge detection technique PSNR (peak signal-to-noise ratio) is being employed. To compute the PNSR value, manual methods have been employed to generate the corresponding ground truths for roof edges. Figure 11 shows a bar graph of different PNSR values for the outcomes of the gradient maps computed via Roberts, Prewitt, Scharr and Sobel operator on the feature map of VGG network separately.

PNSR values for every case are very close to each other. We have experimented with different size and different resolution of images and computed PNSR values but still we got comparable results that are minor difference in PNSR values. This observation made it clear that basically this technique is independent of the edge operator and its performance solely depends upon the computed feature map of the

Fig. 7 Final outcome-VGG followed by Sobel

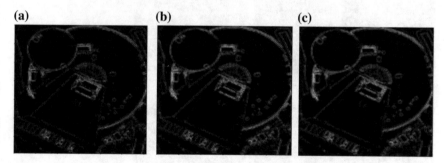

Fig. 8 **a** Final outcome-VGG followed by Roberts, **b** final outcome-VGG followed by Prewitt, **c** final outcome-VGG followed by Scharr

VGG CNN network. Basically, it is the feature map of CNN network that plays the vital role in deciding the final outcome of the proposed technique.

(a) Input Test Images

(b) Gradient Maps of first 3 layers with Prewitt Operator

(c) Gradient Maps of first 3 layers with Sobel Operator

(d) Gradient Maps of first 3 layers with Prewitt Operator

(e) Gradient Maps of first 3 layers with Sobel Operator

Fig. 9 Depiction of gradient maps of first three layers for various test image

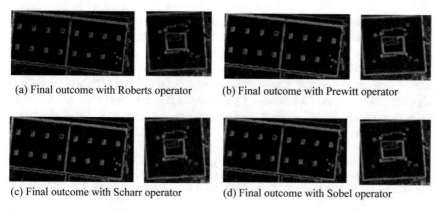

(a) Final outcome with Roberts operator (b) Final outcome with Prewitt operator

(c) Final outcome with Scharr operator (d) Final outcome with Sobel operator

Fig. 10 Final outcome of test images with different operators for variable size images

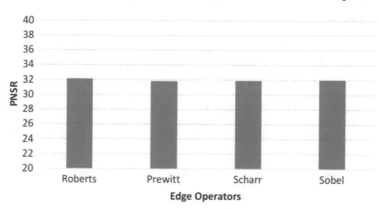

Fig. 11 Bar graph for PNSR of gradient maps for different operators

4 Conclusion

A powerful algorithm is presented for the detection of edges in images using Convolution Neural Networks. This technique can detect edges of roof tops irrespective of the shape of the roof or size of the image. CNN method with transfer learning is employed in this paper to detect the edges because unlike traditional approaches for edge detection its efficiency does not degrade with the increase of noise in images. Traditional edge detection approaches are computationally expensive and non-reliable in case of noise variant image dataset. CNN, on the other hand is very simple, does not require manual feature extraction which makes it fast and CNN can process images of variable size without any additional training. We have employed VGG convolutional neural network for the computation of features which is followed by the implementation of edge detectors to find out the gradient map of the image. The outcomes of this technique depicted better performance for every type of roof images irrespective of its size, location and type of the roof as compared to classical methods.

In future, we are focused towards the implementation of Generative Adversarial Nets (GANs) in this system for edge restoration in images in order to detect the edges more efficiently. Edge restoration application will be extremely beneficial prior to edge detection as it would immensely increase the performance of the system by the restoration of any missed edges in roof images due to presence of a tree, shadow or any obstacle etc.

Acknowledgements This research was financially supported by the Ministry of SMEs and Startups (MSS), Korea, under the "Regional Specialized Industry Development Program (R&D or non-R&D, Project number)" supervised by the Korea Institute for Advancement of Technology (KIAT).

References

1. Wang, R.: Edge detection using convolutional neural network. In: International Symposium on Neural Networks, Advances in Neural Networks—ISNN (20 December 2016)
2. Peli, T., Malah, D.: A study of edge detection algorithms. IEEE Trans. Comput. Graph. Image Process. **20**(1), 1–21 (1982)
3. El-Sayed, M.A., Estaitia, Y.A., Khafagy, M.A.: Automated edge detection using convolutional neural network. Int. J. Adv. Comput. Sci. Appl. (IJACSA) **4**(10) (2013)
4. Torre, V., Poggio, T.A.: On edge detection. IEEE Trans. Pattern Anal. Mach. Intell. **PAMI-8**(2) (March 1986)
5. Marr, D., Hildreth, E.: Theory of edge detection. Proc. R. Soc. London B **207**, 187–217 (1980)
6. Canny, L.: A computational approach to edge detection. IEEE Trans. Pattern Anal. Mach Intell. **8**(1), 679–698 (1986)
7. Rosenblatt, F.: The perceptron: a probabilistic model for information storage and organization in the brain. Psychol. Rev. **65**(6), 386–408 (1958) (Cornell Aeronautical Laboratory)
8. Simonyan, K., Zisserman, A.: Very deep convolutional networks for large-scale image recognition. In: International Conference on Learning Representations (ICLR) (2015)
9. Ahmed, A., Byun, Y.-C.: Edge detection using CNN for roof images. In: The 2019 Asia Pacific Information Technology Conference (January 2019)
10. Xie, S., Tu, Z.: Holistically-Nested Edge Detection. International Journal of Computer Vision 125 (1-3):3-18 (2017)
11. Bertasius, G., Shi, J., Torresani, L.: DeepEdge: A multi-scale bifurcated deep network for top-down contour detection. In: IEEE CVPR 4380–4389 (2015)
12. Shen, W., Wang, Y., Bai, X., Zhang, Z.: Deep-Contour: A deep convolutional feature learned by positive-sharing loss for contour detection. In IEEE CVPR 3982–3991 (2015)
13. Liu, Y.: Richer convolutional features for edge detection. In IEEE Transactions on Pattern Analysis and Machine Intelligence (October 2018)
14. Alom, Md.Z., Taha, T.M., Yakopcic, C., Westberg, S., Hasan, M., Van Esesn, B.C., Awwal, A.A., Asari, V.K.: The history began from AlexNet: a comprehensive survey on deep learning approaches, (March 2018)
15. https://www.kaggle.com/canonwu/roof-images

Improvement of Incremental Hierarchical Clustering Algorithm by Re-insertion

Kakeru Narita, Teruhisa Hochin, Yoshihiro Hayashi and Hiroki Nomiya

Abstract Clustering is employed in various fields for analysis and classification. However, the conventional clustering method does not consider changing data. Therefore, in case of change in data, the entire dataset must be re-clustered. A clustering method has been proposed to update the clustering result obtained by a hierarchical clustering method without re-clustering when a point is inserted by using the center and the radius of a cluster. This paper improves this incremental clustering method. By examining the cluster multimodality which is the property of a cluster having several modes, we can select some points of a different distribution inferred from a dendrogram, and transfer the points in the cluster to a different cluster. In addition, when the number of clusters increases, data points previously inserted are updated by re-insertion. Compared with the conventional method, the experimental results demonstrate that the execution time of the proposed method is significantly less and clustering accuracy is comparable for some data.

Keywords Incremental clustering · Hierarchical clustering · Data insertion · Re-insertion

K. Narita (✉)
Graduate School of Information Science, Kyoto Institute of Technology, Kyoto, Japan
e-mail: m8622036@edu.kit.ac.jp

T. Hochin · H. Nomiya
Information and Human Sciences, Kyoto Institute of Technology, Kyoto, Japan
e-mail: hochin@kit.ac.jp

H. Nomiya
e-mail: nomiya@kit.ac.jp

Y. Hayashi
Research and Development Department, NITTO SEIKO CO., LTD, Ayabe, Japan
e-mail: hayashi@nittoseiko.com

© Springer Nature Switzerland AG 2020
R. Lee (ed.), *Computational Science/Intelligence and Applied Informatics*,
Studies in Computational Intelligence 848,
https://doi.org/10.1007/978-3-030-25225-0_8

1 Introduction

Advances in computer and network technologies allow us to obtain various information and services via the Internet, e.g., cloud computing [1]. Cloud computing is not dependent on fixed terminals, and is expected to enable handling of considerable data easily. In addition, the Internet of Things (IoT) technology, which can be used to control and obtain data from different devices, is expected to drive change [5, 11]. For example, IoT is changing manufacturing industries. In cloud manufacturing [9], manufacturing equipment is connected and controlled via the Internet, and considerable data are transmitted over the Internet [1, 11]. Machine learning can be employed to derive useful information from such large amounts of data [12, 15].

Machine learning can be classified as supervised and unsupervised [12, 15]. Supervised learning involves labeled input-output pairs. A supervised learning algorithm produces an inferring function to estimate an output from an input. In contrast, unsupervised learning attempts to find the underlying structure in a set of data points. Note that labeled data are not required for unsupervised learning; thus, unsupervised learning is useful when labeled training data are unavailable. In this paper, we focus on clustering, which is an important unsupervised learning technique.

Clustering, which is used for the analysis and classification of various data, involves generating groups, i.e., clusters, that have similar characteristics. However, the conventional clustering method does not consider changes in the data. Therefore, when new data are inserted, all data must be re-clustered, which is crucial in dynamic environments. When the amount of updating is small, the classification result does not change significantly. However, re-clustering requires considerable time; therefore, a method to partially change clusters without re-clustering when only a small amount of data is inserted is required.

Several incremental clustering methods have been proposed previously [2–4, 6], which attempt to update a single cluster or a few clusters locally when a data point is inserted, and such methods attempt to store the inserted data point in a cluster that has maximum similarity to it. Although these methods have achieved good performance, they are based on non-hierarchical clustering methods and their application is quite limited. Ribert et al. proposed an incremental hierarchical clustering method [14] that attempts to find the best insertion point for a new data item. With this method, memory cost is very low; however, the number of computation of distances between clusters does not decrease significantly. Thus, a method that involves the computation of fewer distances is required. Gurrutxaga et al. proposed another incremental hierarchical method [7]. This method reduced time complexity, but accuracy is worse than the conventional method in some cases. Therefore, improving clustering accuracy is required. Narita et al. proposed yet another incremental clustering method [13]. The center and the radius of the cluster are used to determine the cluster and location into which new data will be inserted. In addition, their method employed the concept of outliers and considers the creation of a new cluster due to data insertion. However, accuracy is worse than that of the conventional method. Improving cluster accuracy is required.

In this study, we attempt to improve the incremental clustering method proposed by Narita et al. [13]. In the improved method, we move some points in a given cluster to a different cluster by determining if the given cluster is multimodal. In addition, when the number of clusters increases, the previously inserted points are updated by re-insertion. We compare the proposed and conventional methods using several datasets, and the experimental results demonstrate that, compared with the existing method's re-clustering time, the proposed method can classify points faster when inserting new points.

The remainder of the paper is organized as follows. Section 2 describes work related to incremental clustering methods. Section 3 proposes an improved incremental clustering method. Section 4 compares the proposed method with Ward's method [10] relative to the execution time, concordance rate, and overall accuracy. Section 5 discusses the result. Finally, Sect. 6 provides the conclusions and suggestions for future work.

2 Related Work

Ester et al. [4] proposed an incremental clustering algorithm based on DBSCAN (Density-Based Spatial Clustering of Applications with Noise). Due to the DBSCAN's density-based nature, the insertion or deletion of an object only affects a small neighborhood of objects; therefore, by changing only the neighborhood, efficient algorithms can be applied for incremental insertions and deletions to an existing clustering result. The method proposed by Ester et al. [4] yields the same result as that of DBSCAN, and the experimental results demonstrate that incremental algorithm is faster than DBSCAN.

Charikar et al. [3] proposed an efficient incremental clustering model for a dynamic environment to maintain small cluster diameters when new points are inserted. They analyzed several greedy algorithms and demonstrated that such algorithms perform poorly. They proposed deterministic and randomized incremental clustering algorithms, i.e., a doubling algorithm and a clique partition algorithm. The doubling algorithm creates a cluster-centric graph and combines randomly selected clusters and their neighborhoods. The clique partition algorithm creates a cluster-centric graph and combines clusters by solving the minimum clique partitioning problem. Charikar et al. showed that these algorithms demonstrate provably good performance.

Can [2] proposed the incremental algorithm C2ICM with improved C3M to address problems associated with regularly updated document databases. *Seed power* was employed in this method to determine the cluster to which new documents belong. Here, seed power is calculated using cover-coefficient in C3M. The C2ICM algorithm computes the number of clusters and seed powers of documents in the updated document database and selects the cluster seeds. Then, to cluster these documents by assigning them to the cluster of the seed that covers them the most and any document not covered by any seed are grouped into a ragbag cluster. The experimental results demonstrate that C2ICM is faster than C3M and yields the same result.

Gupta et al. [6] proposed a K-means clustering-based incremental clustering algorithm that employs threshold value. If the distance between a new point and the existing cluster center is greater than the given threshold, a new cluster center is formed and the point becomes the new cluster centroid; otherwise, the point is inserted into the existing cluster, and its centroid is updated. Note that this method does not need the information on the number of clusters. They demonstrated that experiments can be conducted in less time as compared to the conventional k-means algorithm.

Ribert et al. [14] proposed an incremental clustering method to update the hierarchical representation to address the problem associated with the hierarchical method's memory requirements. Here, the concept of the region of influence is introduced to determine the position of the new element in the hierarchy. With this algorithm, clustering is performed on places other than the unchanged taxonomy. Their experimental results demonstrated that memory cost was reduced significantly.

Gurrutxaga et al. [7] proposed an incremental hierarchical clustering method to reduce time complexity. They decided to insert points by searching the dendrogram from its root node recursively. They proved that time complexity could be reduced and showed that their experimental results give better results than the conventional method in some cases.

Narita et al. [13] proposed an incremental hierarchical clustering method. They defined two features, the center and the radius of the cluster, and use these features to determine the cluster and location into which new data will be inserted. In addition, their method employed the concept of outliers and considers the creation of a new cluster due to data insertion. A cluster radius may be enlarged by inserting points. If distances between the cluster and outliers are close, the cluster absorbs outliers. If the cluster becomes close to other cluster, two clusters are combined. In inserting and combining process, the inserting location is determined by two features. However, updated dendrogram is created by using the similarity based on the conventional method. Distances to be compared are different. In addition, there is a problem that some different groups may exist in the same cluster.

The differences between these existing methods and the proposed method are summarized as follows. First, some of these existing methods [2–4, 6] are based on nonhierarchical clustering. Unlike the method proposed by Charikar et al. [3], the proposed method does not require the number of clusters. Similar to the method proposed by Gupta et al. [6], the proposed method does not require all points to be allocated to a cluster. The method proposed by Can [2] focuses on document databases. In contrast, the proposed method handles numerical data. The hierarchical clustering method proposed by Ribert et al. [14] reduced memory usage, whereas the proposed method reduces execution time. The result of the hierarchical clustering method proposed by Gurrutxaga et al. [7] is different from that of the conventional method, whereas we attempt to achieve the same result as the conventional method.

The method proposed in this paper addresses to Narita et al.'s problem [13].

3   Proposed Method

We follow the Narita et al.'s method. In order to increase the clustering accuracy, we improve their clustering method. We changed the processings of "Insertion" and "Combination," and added new processings: re-insertion of points, and division of a cluster.

3.1   Inserting a Point into a Cluster

Let x_{insert} be a new point. If x_{insert} is inside any cluster, C_{near} is expressed as follows:

$$C_{near} = \underset{C_i}{\arg\min} \left(\{ a \mid a = d(\bar{x}_{C_i}, x_{insert}), a \leq r_{C_i} \} \right) \tag{1}$$

where C_i is the i-th cluster ($0 \leq i \leq NC$), NC is the number of clusters, $\bar{x}_{C_i}$ is the center of C_i, r_{C_i} is the radius of C_i, and d(x,y) is an arbitrary distance function between x and y.

When x_{insert} is outside the i-th cluster C_i, the distance between point x_{insert} and cluster C_i is calculated using Eq. (2). As described in Eq. (3), C'_{near} is the cluster with the smallest distance to x_{insert}.

$$d'_{C_i} = d(\bar{x}_{C_i}, x_{insert}) - r_{C_i} \tag{2}$$

$$C'_{near} = \underset{C_i}{\arg\min} (d'_{C_i}) \tag{3}$$

After determining the cluster to which x_{insert} belongs, the proposed method determines where to insert it into a partial cluster by tracing a dendrogram as follows.

[Insertion Process]
1. Let C_{now} denote the the selected cluster.
2. Calculate the similarity between x_{insert} and C_{now}.
3. If the similarity is greater than that of C_{now}, a partial cluster is formed with x_{insert} and C_{now}, and the process terminates.
4. Otherwise, calculate the similarities of the children of C_{now}. Then let C_{now} be the one with the smallest similarity.
5. Repeat Steps (2)–(4) until C_{now} reaches to leaves.

Note that the algorithm uses the similarity calculation method employed in the original clustering method. The centers and the radii of the visited partial clusters are updated in the same manner as [13].

3.2   Combining Clusters

Note that the cluster creation process may result in many small clusters. In addition, the cluster size will increase as points are inserted. However, it may be better to combine adjacent clusters and generate a single large cluster. We improve combining algorithm to achieve better results than previous method. It is summarized as follows.

[Combination Process]
1. Find the minimum distance from the center of the selected cluster to the center of all clusters.
2. If the distance is less than or equal to the sum of two cluster radii, the process terminates; otherwise, calculate a pseudo-F value [15] ($F_{current}$) for the current clustering result.
3. If two clusters are combined, calculate a pseudo-F value ($F_{combine}$) for the combined cluster.
4. If $F_{combine}$ is greater than $F_{current}$, search to combine with the partial cluster using the insertion process.

3.3   Re-inserting Points

When the number of clusters increases due to the creation or division process, points previously inserted to an existing cluster may become closer to the new cluster than the current cluster. Therefore, if the number of clusters increases, points inserted into each close cluster must be re-inserted. Once the points have been re-inserted, they are never re-inserted. The re-insertion process is performed when a cluster is created or divided. The points to be re-inserted are "inserted points after fixed" and "points not belonging to the newly created cluster" or "points not belonging to the cluster prior to division." If the points are re-inserted prior to removing all target points, the points to be removed may affect the re-insertion. Therefore, all target points should be removed and re-inserted. The re-insertion algorithm is as follows.

[Re-insertion Process]
1. Remove all target points from clusters.
2. Perform Steps (3) and (4) in the order of insertion before.
3. Insert the removed points into the closest cluster or identify removed points as outliers using the processes of "Inserting a Point into a Cluster" (described in Sect. 3.1) and "Determining a Point as an Outlier" (in [13]).
4. If the removed point is in a set of outliers, delete it from the cluster.
5. Fix the clustering result.

3.4 Dividing a Cluster

An existing cluster will expand when points are inserted. However, if biased points are inserted into the cluster, it may be better to divide the cluster. The process of dividing a cluster into two clusters by inserting points is described as follows.

Assume that the cluster to be divided has multiple peaks, because a cluster is a set of similar points. Therefore, in the cluster, the multimodality, the property in distribution of having several modes, is evaluated using the DIP test [8]. Here, let the data used for the DIP test be the distances between a pseudo-mode value and points in the cluster. The pseudo-mode value is defined as a vector comprising the mode of each dimension in the cluster.

A cluster with multimodality is divided, and a node corresponding to the cluster is divided into two child nodes in a dendrogram. Then, pseudo-F values are used to determine whether to adopt the division result. The division process is performed when clusters are not combined. The division process is summarized as follows.

[Division Process]
1. Perform a multimodal test with 5% significance level for a cluster into which a point is inserted.
2. If the cluster has multimodality, calculate a pseudo-F value.
3. Calculate a pseudo-F value when the cluster is divided in two using a dendrogram.
4. After division, if the pseudo-F value is greater than the value prior to division, adopt the division result. Otherwise, the cluster is not divided.

4 Experiment

4.1 Experimental Method

4.1.1 Purpose

To demonstrate the proposed method's performance, we compared its execution time with that of the conventional clustering method (Ward's method) and that of the previous method [13]. In addition, we compared the clustering results of the conventional method with those of the proposed method to demonstrate its concordance rate.

4.1.2 Data

Five datasets were used in the experiment.
Data 1 was a synthetic data set comprising the followings.

Table 1 Detail of correct label in Data 2–5

Cluster label	Material	Crack
1	A	None
2	A	Small
3	A	Large
4	B	Large

- 1800 initial points.
  600 points were generated from each of the three kinds of 2D normal distributions with different mean. All variances were 1.0, and the means were $(0, 0)$, $(8, 0)$, and $(8, 8)$.
- 360 insertion points.
  300 random points were generated from the 2D normal distribution with mean $(0, 8)$ and variance 1.0, and 60 2D uniform random points were generated in the interval $[-2, 10)$.

Data 2–5 have the following feature values.

- Calculate the average values of the frequency components of the power spectrum obtained from 1024 FFTs at intervals of 128 samples of the same vibration waveform.
- Principal component values of 513 points were obtained using principal component analysis and then arranged in ascending order.

We used 3, 12, and 30 dimensions with cumulative contribution rates exceeding 0.4, 0.7, and 0.8, and 20 dimensions between 0.7 and 0.8. Data 2–5 were set to 3-, 12-, 20-, and 30-dimensional data, respectively. These data can be clustered into four groups (Table 1).

4.1.3   Method

The previous method is only evaluated with data 1, while the conventional method is evaluated with all data sets.

First, initial points were clustered by using Ward's method, and the radii and centers of all clusters were calculated. Here, for the proposed method, we measured the time required to insert a point. In contrast, for the conventional method, the re-clustering time was measured.

The cluster obtained by the proposed method, which has the most matching points with a cluster of the re-clustering result, was considered the same cluster. The concordance rate (Eq. 4) was used to measure how well the result agrees with that obtained by the conventional method. In addition, if data have the correct label, the overall accuracy defined by Eq. (5) indicates how correct the result is.

$$Concordance\ rate = \frac{Total\ number\ of\ points\ in\ the\ same\ cluster}{Number\ of\ all\ points} \qquad (4)$$

$$Overall\ accuracy = \frac{Number\ of\ points\ classified\ in\ the\ correct\ cluster}{Number\ of\ all\ points} \qquad (5)$$

The threshold $thr_{outlier}$ (number of outliers) was varied from 10 to 30 in increments of 5, and the threshold $thr_{cluster}$ (number of points in the cluster) was varied from 20 to 40 in increments of 5.

4.1.4 Environment

The experimental environment is summarized as follows.

- A Dell OptoPlex 7050 PC (CPU 3.60 GHz, RAM 8.00 GB, OS Windows 10 Pro) was used.
- Time was measured using Python's time function perf_counter().
- The program was written in Python (3.6.1) with NumPy (1.13 + mkl) and SciPy (0.19.1).

4.2 Experimental Results

For Data 1, the minimum and maximum average execution times per insertion of the proposed method and the previous method, and the re-clustering time of the conventional method are shown in Fig. 1. Note that the maximum and minimum average execution times of the proposed method are nearly the same; therefore, only the maximum average execution times per insertion of the proposed method and the re-clustering time of the conventional method for Data 2–5 are shown in Figs. 2, 3, 4 and 5.

As shown in Fig. 6, the concordance rate demonstrated two tendencies with Data 1: for $thr_{outlier} = 20\ (30)$ and $thr_{cluster} = 40\ (35)$. The concordance rate tendencies for Data 2–5 are shown in Figs. 7, 8, 9 and 10. These tendencies were obtained for $thr_{outlier} = 20$ and $thr_{cluster} = 20$.

Data 2–5 had correct labels, and were grouped into four clusters. Tables 2 and 3 show the results obtained based on a pseudo-F value for the proposed and the conventional methods, respectively. With pseudo-F values, the correct answer and the number of clusters differ; thus, Table 4 shows a comparative result when the dendrogram was cut such that the number of clusters was four after classification.

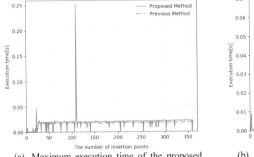

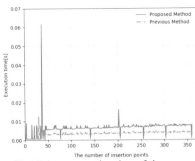

(a) Maximum execution time of the proposed method and the previous method

(b) Minimum execution time of the proposed method and the previous method

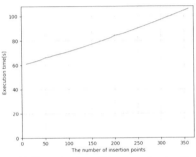

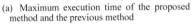

(c) Execution time of the conventional method

Fig. 1 Execution time result with Data 1

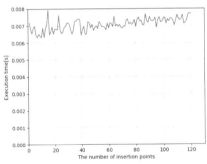

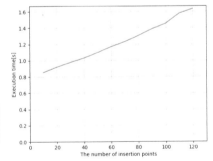

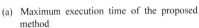

(a) Maximum execution time of the proposed method

(b) Execution time of the conventional method

Fig. 2 Execution time result with Data 2

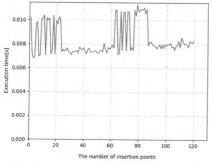 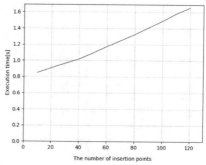

(a) Maximum execution time of the proposed method

(b) Execution time of the conventional method

Fig. 3 Execution time result with Data 3

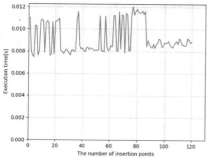 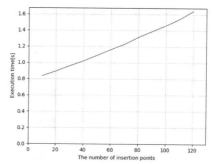

(a) Maximum execution time of the proposed method

(b) Execution time of the conventional method

Fig. 4 Execution time result with Data 4

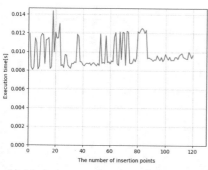 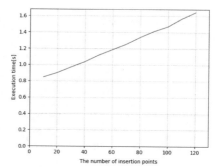

(a) Maximum execution time of the proposed method

(b) Execution time of the conventional method

Fig. 5 Execution time result with Data 5

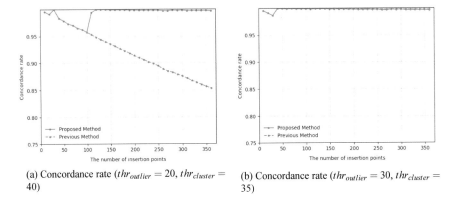

(a) Concordance rate ($thr_{outlier} = 20, thr_{cluster} = 40$)

(b) Concordance rate ($thr_{outlier} = 30, thr_{cluster} = 35$)

Fig. 6 Data 1 concordance rate of the proposed method and the previous method

Fig. 7 Data 2 concordance rate

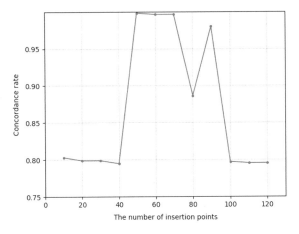

Fig. 8 Data 3 concordance rate

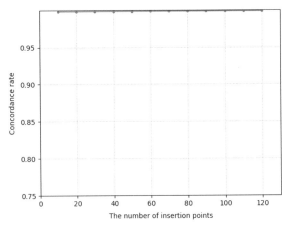

Fig. 9 Data 4 concordance rate

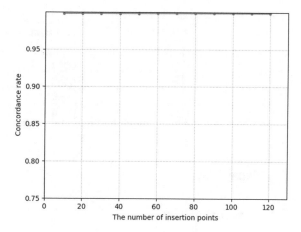

Fig. 10 Data 5 concordance rate

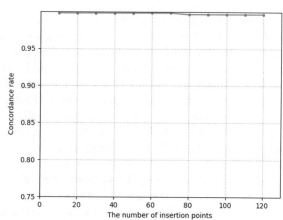

Table 2 Clustering by the proposed method based on Pseudo-F

Data ID	Dimension	Number of clusters	Overall accuracy
2	3	4	90.6%
3	12	2	–
4	20	2	–
5	30	2	–

Table 3 Clustering by the conventional method based on Pseudo-F

Data ID	Dimension	Number of clusters	Overall accuracy
2	3	3	–
3	12	2	–
4	20	2	–
5	30	2	–

Table 4 Overall accuracy (four clusters)

Data ID	Dimension	Proposed method (%)	Conventional method (%)
2	3	90.6	98.6
3	12	96.9	96.7
4	20	98.8	96.7
5	30	91.0	98.4

5 Discussion

5.1 Execution Time

As shown in Fig. 1a, a definite peak is evident because the re-insertion process involves an increased number of clusters, which consumes time. In practice, the re-insertion process may be unlikely. Therefore, the result needs to be fixed by another process. As can be seen by comparing Figs. 1a, b, the execution times differ even if the peaks are excluded. This difference may be caused by whether pseudo-F values are calculated for the cluster combination process. Compared with the previous method, the execution time of the proposed method slightly increases except for the peaks. This increase is due to the addition of division process. In the remaining figures for the proposed method, these times are nearly the same irrespective of thresholds $thr_{cluster}$ and $thr_{outlier}$. This may be because outliers were not processed frequently.

The processing time of the proposed method is obviously less than the re-clustering time of the conventional method. Thus, the proposed method is superior in terms of time cost. The conventional clustering method calculates distances for all pairs of all points for re-clustering, while the proposed method calculates distances between pairs of a few points. As the both time complexity is $O(n^2)$, the proposed method successfully reduces the calculation time.

5.2 Concordance Rate

The concordance rates for Data 1 differ according to the parameters because the arguments relate to outliers, and such arguments affect data that cause outliers. As shown in Fig. 6a, the concordance rate of the previous method is degraded as points are inserted. While in the proposed method, although the concordance rates first degrade temporarily as in the previous method, if sufficient data are available, it is improved and it is close to 1. In Fig. 6b, both methods give almost the same concordance rate. For Data 2, the concordance rate was reduced due to the difference between the number of clusters obtained by re-clustering and that of the clusters obtained by the proposed method. Note that high concordance rates were maintained for Data 3–5.

5.3   Overall Accuracy

Tables 2 and 3 show that only the proposed method with Data 2 is able to obtain the correct number of clusters by pseudo-F values. For the other data, the numbers of clusters differed. For Data 3, 4, and 5, dataset is roughly classified by material differences rather than cracks. Note that the overall accuracy was calculated only when the number of clusters was the correct number of clusters.

As seen in Table 4, both methods obtained high overall accuracy (greater than 90%) for all data. For Data 2 and 5 with the proposed method, accuracy for cluster label 3 is low; however, for the other cluster labels, it was as high as the other data. The best overall accuracy obtained by the proposed method was obtained with Data 4. This indicates that high accuracy can be achieved if the correct number of clusters is obtained using any index. We expect that employing other index to determine the number of clusters may lead to better results.

5.4   Thresholds

The execution time varies depending on thresholds $thr_{cluster}$ and $thr_{outlier}$, and this may depend on the points used. Note that this may affect the concordance rate. An outlier tends to occur when $thr_{cluster}$ is small and $thr_{outliers}$ is large.

6   Conclusion

In this paper, we improved the incremental clustering method proposed by Narita et al. [13]. We divided a given cluster into two by determining whether the given cluster has multimodality. In addition, when the number of clusters increases, the previously inserted points are re-inserted. The experimental results demonstrated that the proposed method requires less execution time than Ward's method. The concordance rate of the proposed method sometimes degraded due to the difference in the number of clusters, while a high concordance rate was achieved with some data, and it was greatly improved compared with the previous method. In addition, overall accuracy of proposed method was high if the correct number of clusters was obtained. To obtain the correct number of clusters, it could be better to employ an indicator other than the pseudo-F.

Note that we focused on inserting new points. In the future, we plan to examine the deletion of unnecessary points. In addition, we plan to evaluate the proposed method using other practical data.

Acknowledgements We are deeply grateful to Mr. Masakazu Ishihara from NITTO SEIKO CO., LTD., who provided us valuable data and discussed them eagerly.

References

1. Agrawal, D., Das, S., El Abbadi, A.: Big data and cloud computing: current state and future opportunities (2011).https://doi.org/10.1145/1951365.1951432
2. Can, F.: Incremental clustering for dynamic information processing. ACM Trans. Inf. Syst. **11**(2), 143–164 (1993). https://doi.org/10.1145/130226.134466
3. Charikar, M., Chekuri, C., Feder, T., Motwani, R.: Incremental clustering and dynamic information retrieval. SIAM J. Comput. **33**(6), 1417–1440 (2004). https://doi.org/10.1137/S0097539702418498
4. Ester, M., Kriegel, H.P., Sander, J., Wimmer, M., Xu, X.: Incremental clustering for mining in a data warehousing environment. In: Proceedings of the 24th International Conference on Very Large Data Bases, VLDB '98, pp. 323–333. San Francisco, CA, USA (1998). http://dl.acm.org/citation.cfm?id=645924.671201
5. Gubbi, J., Buyya, R., Marusic, S., Palaniswami, M.: Internet of things (IoT): a vision, architectural elements, and future directions. Future Gener. Comput. Syst. **29**(7), 1645–1660 (2013). https://doi.org/10.1016/j.future.2013.01.010, http://www.sciencedirect.com/science/article/pii/S0167739X13000241
6. Gupta, N., Ujjwal, R.L.: An efficient incremental clustering algorithm. World Comput. Sci. Inf. Technol. J. **3**(5), 97–99 (2013)
7. Gurrutxaga, I., Arbelaitz, O., Ignacio Martín, J., Muguerza, J., Pérez, J., Perona, I.: Sihc: a stable incremental hierarchical clustering algorithm, pp. 300–304 (2009)
8. Hartigan, J.A., Hartigan, P.M.: The dip test of unimodality. Ann. Statist. **13**(1), 70–84 (1985).https://doi.org/10.1214/aos/1176346577
9. He, W., Xu, L.: A state-of-the-art survey of cloud manufacturing. Int. J. Comput. Integr. Manuf. **28**(3), 239–250 (2015). https://doi.org/10.1080/0951192X.2013.874595
10. Ward Jr., J.H.: Hierarchical grouping to optimize an objective function. J. Am. Stat. Assoc. **58**(301), 236–244 (1963). https://doi.org/10.1080/01621459.1963.10500845
11. Lee, I., Lee, K.: The internet of things (IoT): applications, investments, and challenges for enterprises. Bus. Horiz. **58**(4), 431–440 (2015). https://doi.org/10.1016/j.bushor.2015.03.008, http://www.sciencedirect.com/science/article/pii/S0007681315000373
12. Marsland, S.: Machine Learning: An Algorithmic Perspective, 1st edn. Chapman & Hall/CRC (2009)
13. Narita, K., Hochin, T., Nomiya, H.: Incremental clustering for hierarchical clustering. In: Proceedings of 5th International Conference on Computational Science/Intelligence and Applied Informatics (CSII 2018), pp. 102–107 (2018). https://doi.org/10.1109/CSII.2018.00025
14. Ribert, A., Ennaji, A., Lecourtier, Y.: An incremental hierarchical clustering. In: Proceedings of 1999 Vision Interface Conference, pp. 586–591 (1999)
15. Zumel, N., Mount, J.: Practical data science with R. Manning (2014)

A New Probabilistic Tree Expression for Probabilistic Model Building Genetic Programming

Daichi Kumoyama, Yoshiko Hanada and Keiko Ono

Abstract This paper proposes a new expression of probabilistic tree for probabilistic model building GPs (PMBGP). Tree-structured PMBGPs estimate the probability of appearance of symbols at each node of the tree from past search information, and decide the symbol based on the probability at each node in generating a solution. The probabilistic tree is a key component of PMBGPs to keep appearance frequencies of symbols at each node by probabilistic tables, and the probabilistic prototype tree (PPT) expressed as a perfect tree has been often employed in order to include any breadth of trees. The depth of PPT is an important parameter that involves trade-off between the search accuracy and the computational cost. The appropriate depth depends on problems and is difficult to be estimated in advance. Our proposed probabilistic tree is constructed by a union operator between probabilistic trees, and its depth and breadth are extendable, so that the depth of the probabilistic tree is not required as a parameter and it needs not necessarily be a perfect tree. Through numerical experiments, we show the effectiveness of the proposed probabilistic tree by incorporating it to a local search-based crossover in symbolic regression problems.

1 Introduction

Genetic programming (GP) is one of promising program evolution algorithms based on the evolutionary computation. Similar to the search framework of genetic algorithms (GA), most GPs perform a selection, a crossover and a mutation to develop a program. In tree-structured GPs, the dependencies between nodes in a tree have

D. Kumoyama
Graduate School of Science and Engineering, Kansai University, Osaka, Japan

Y. Hanada (✉)
Faculty of Engineering Science, Kansai University, Osaka, Japan
e-mail: hanada@kansai-u.ac.jp

K. Ono
Department of Electronics and Informatics, Ryukoku University, Shiga, Japan
e-mail: kono@rins.ryukoku.ac.jp

© Springer Nature Switzerland AG 2020
R. Lee (ed.), *Computational Science/Intelligence and Applied Informatics*,
Studies in Computational Intelligence 848,
https://doi.org/10.1007/978-3-030-25225-0_9

a big influence on the evaluation value of an individual. Due to this, it is important for the crossover to generate offspring by building and keeping preferable traits considering the relationship of locations of nodes. However, good traits differ depending on problems, and it is often difficult to know in advance what traits should be built. To improve search abilities of GP, dependencies between nodes or semantics of subtrees have been considered, based on symbolic information, in reproduction operators. Among them, probabilistic model building GP (PMBGP) is one of promising approaches to help estimating good traits [1–5]. PMBGPs estimate the preferable traits by using probability of appearance of symbols at each node of the tree from past search information, and decide the symbol based on the probability at each node in generating solutions.

In our previous work, we incorporated the probabilistic prototype tree (PPT), which is used in one of PMBGPs [1, 3, 5, 6], to a local search-based crossover, deterministic multistep crossover fusion (dMSXF) [7]. PPT is a perfect N-ary tree in which each node has a probabilistic table defined by frequencies of symbols obtained by past search information. N is determined by the problem definition and set as the maximum number of operands (child nodes) that an operator (parent node) should have. As a probabilistic model, we adopted the probabilistic table considering a dependency relationship between the parent node and the child node [6], and showed that the generation of preferable traits was enhanced in neighborhood solutions of dMSXF. However, probabilistic model building approaches often involve a computational cost issue. The depth d of tree is one of important parameters of PPT, and it is desired to be set deep enough in order to include any depth of solution generated in the search. A high search performance is expected under a large d, but the number of probabilistic tables that PPT keeps is $N^d - 1$, so that a larger d requires a larger memory and causes larger calculation cost. Due to computational cost, PPT sometimes could not be set to large enough to the problem, and it is difficult to improve the search performance depending on the problem. Moreover, it is also difficult to estimate an appropriate depth to an individual problem, in advance.

In this paper, we propose a new expression of probabilistic tree with the extendable depth and the breadth. We introduce a union operator between probabilistic trees to merge their probabilistic information, so that the depth of the probabilistic tree is not required as a parameter and it needs not necessarily be a perfect tree. To show the effectiveness of the probabilistic tree, we incorporate it to the crossover, dMSXF, and evaluate the search performance in symbolic regression problems.

2   A New Expression of Probabilistic Tree

2.1   PPT and Parent-Child Probabilistic Model

PMBGPs estimate the probability of appearance of symbols at each node from past search information, and decide the symbol based on the probability at each node in

Fig. 1 An example of PPT with $N = 2$ and $d = 4$ with parent-child model in the symbolic regression problem with the nonterminal set $V^{NT} = \{+, -, *, /\}$ and the terminal set $V^T = \{x, 1\}$

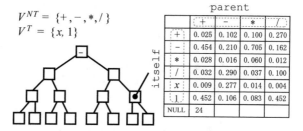

$$V^{NT} = \{+, -, *, /\}$$
$$V^T = \{x, 1\}$$

parent

itself	+	−	*	/
+	0.025	0.102	0.100	0.270
−	0.454	0.210	0.705	0.162
*	0.028	0.016	0.060	0.012
/	0.032	0.290	0.037	0.100
x	0.009	0.277	0.014	0.004
1	0.452	0.106	0.083	0.452
NULL	24			

generating a solution [1, 3, 5, 6]. To keep an appearance frequency of each symbol at a node, the probabilistic prototype tree (PPT) has been often used in PMBGPs. PPT is expressed as a perfect tree, and each node in PPT keeps a probability table that is relevant to all defined symbols.

A type of probability table depends on the employed probabilistic model. In our previous work, we adopted the parent-child model used in Estimation of Distribution Programming (EDP) [6]. Figure 1 illustrates a PPT with the parent-child model. This example is an instance of symbolic regression problem with arithmetic operators, and PPT is presented as a perfect binary tree. Each node keeps a probability table that is relevant to symbols of parent and child, and all values of probability table are initially set to 0. In Fig. 1, NULL means the frequency where node has not existed at this location. Only for the root node, the probability table keeps symbols frequency at this location.

The procedure of updating or referring to the probability table is as follows: (1) Find the pair of corresponding nodes between a solution tree and PPT based on node's location, then (2) update probability tables by the symbols of solution tree at all corresponding nodes, or deciding a symbol at a node of solution tree stochastically based on the corresponding probability table.

The depth of PPT is a parameter and only the nodes located within the depth can update or refer to probability tables. Information of symbols of nodes that located deeper than the depth are not reflected, and symbols of nodes deeper than the depth are decided randomly during generating a solution.

2.2 Proposal of Union Probabilistic Tree

It is desired to be set deep enough in order to include any depth of solution generated in the search, however PPT with a large depth requires a larger memory and causes larger calculation cost. In addition, it is also difficult to estimate an appropriate depth to an individual problem, in advance. In this paper, we propose a new expression of probabilistic tree named the union probabilistic tree (UPT). UPT is a probabilistic tree with the extendable depth and the breadth. Each node of UPT has a probability table as with PPT, but the depth of UPT is not required as a parameter and it needs not necessarily be a perfect tree.

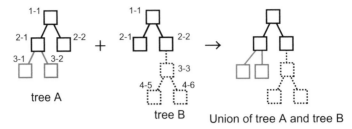

d-#: depth and the location number

Fig. 2 An example of the union operator between two trees (in the case of $N = 2$): First the common tree consisting of the nodes that have the same labels between the tree A and the tree B is exracted (solid lined nodes). Then a new tree is generated by appending all nodes observed only in the tree A or the tree B to the common tree

We first define the union operator between two trees. The procedure of union operation is as follows: (1) Assign a label d-# to each node in both trees where d is the depth and # indicates the location number with counting from leftmost node in each depth on the perfect tree. (2) Make a new tree consisting of only nodes that have the same labels between both trees. (3) Append nodes to the new tree from trees if the same label does not exist between both trees.

Figure 2 illustrates an example of union operator. Note that UPTs are also not perfect trees as with the tree A or the tree B as shown in Fig. 2.

At step (1) of the union operator, the label of the root node is set to 1-1, then we can calculate the location number # of node c as $N^{(\#_p-1)} + \ell$, where $\#_p$ is the location number of the parent of the node c, and ℓ indicates that the node c is the ℓ-th leftmost child.

A solution tree can be converted into the probabilistic tree by keeping its topology and reflecting the symbol information to the initialized probability table of each node. The union operator above can be applied to probabilistic trees, and we define a probabilistic tree constructed by the union operator as UPT. In the union operator between two UPTs, at step (2), the probability table of each node in the new UPT is calculated or updated by adding probability tables of nodes with the same label of both UPTs.

UPT is constructed through runs of GP. UPT is initialized as a probability tree that has only a root node. The UPT is then updated by a solution tree selected in some way from generated solutions in the search.

3 Probabilistic Model Building GP with UPT

3.1 Deterministic Multi-step Crossover Fusion

In this paper, we employ deterministic multistep crossover fusion (dMSXF) [8, 9] as the crossover of GP. In solving combinatorial optimization problems involving complex constraints, local searches have been incorporated to crossover in order to adjust the structural details of solutions, since a simple recombination often brings a drastic change in solutions and might break favorable characteristics. dMSXF is successful one of such a memetic crossover for combinatorial structures.

dMSXF performs a sequence of local search which gradually moves the offspring from its initial point to the other parent. Due to this mechanism, it can generate a wide variety of solution between parents with keeping their traits. The procedure of dMSXF is as follows.

Procedure of dMSXF

Step 1   Let p_1, p_2 be parents and set their offspring $C(p_1, p_2) = \phi$.

Step 2 $k = 1$. Set the initial search point $x_1 = p_1$ and add x_1 into $C(p_1, p_2)$.

Step 3   /Step k/ Prepare $N(x_k)$ composed of μ neighbors generated from the current solution x_k. $\forall y_i \in N(x_k)$ must satisfy $d(y_i, p_2) < d(x_k, p_2)$.

Step 4   Select the best solution y from $N(x_k)$. Let the next search point x_{k+1} be y, and x_{k+1} is added into $C(p_1, p_2)$.

Step 5   Set $k = k + 1$ and go to 2. until $k = k_{max}$ or x_k equals p_2.

At step 3 of the procedure of dMSXF, every neighborhood candidates y_i ($1 \le i \le \mu$) generated from x_k must be closer to p_2 than x_k. dMSXF requires two parameters, k_{max} and μ and $C(p_1, p_2)$ is comprised of $\{x_1, x_2, \ldots, x_{k_{max}}\}$.

3.2 Definition of Distance

To apply dMSXF to a problem, a problem-specific distance measure should be defined. The definition of distance between the tree T_a and the tree T_b is as following.

First, we assign loci to the nodes included in the largest common subtree of T_a and T_b. The largest common subtree is the isomorphic subtree that has the maximum number of vertices between the two trees. For the nodes included in the largest common subtree, let U_a and U_b denote the set of nodes where the locus is same but the symbol is different each other, in respective trees. Let D_a and D_b denote the sets of nodes that are not included in the largest common subtree, in respective trees. The distance between trees T_a and T_b, $d(T_a, T_b)$, is defined as

$$d(T_a, T_b) = |U_a| + |D_a| + |D_b|, \tag{1}$$

where $|\cdot|$ means the number of components included in the set. The detail of the distance is explained in [7].

3.3   Neighborhood Generation Based on Probabilistic Table

The neighborhood solutions approach the other parent by getting its traits little by little with three operations in the local search of dMSXF. In our previous work, we proposed a neighborhood generation based on a replacement, an insertion and a deletion of nodes [7]. Here, we introduce the following six functions and three fundamental operations to explain the neighborhood generation method.

3.3.1   Functions

Most problems to be solved by GP have a constraint in terms of the number of child nodes that nonterminal nodes should keep. For example, the number of operands that each arithmetic operator has is defined in the case of symbolic regression problems. Functions are defined as follows, for describing the operators that handle tree structures.

- $parent(n)$ returns the parent node of the node n.
- $child(n)$ returns the set of child nodes of the node n.
- $st(n)$ returns the subtree whose root node is the node n.
- $arg(n)$ returns the number of child nodes that the node n should keep.
- $symbl(n)$ returns the symbol of the node n.
- $loc(n)$ returns the locus of the node n if it is defined, otherwise returns $null$.

3.3.2   Operations

Neighborhood solutions are generated by three types of operation; *Replace*, *Delete*, *Insert*. All operations work on nodes included in the largest common subtree. Among them, *Delete* and *Insert* refer to the probabilistic tree.

Given two trees T_a and T_b, *Replace* is the operator that replaces the symbol of a node u of T_a with that of a node v of T_b. The operation of *Replace* is as follows.

Replace (u, v)

Step 1   Select one pair of nodes u and v that satisfy $arg(u) = arg(v)$ and are located in the same gene locus, respectively from U_a and U_b.
Step 2   Substitute $symbl(v)$ for $symbl(u)$ in T_a.

Delete is applied when $arg(u) > arg(v)$. This operator deletes a descendent subtree of depth 1 of the node u in T_a as follows.

Delete (u, v)

Step 1   Select one pair of nodes u and v that satisfy $arg(u) > arg(v)$ and are located in the same gene locus, respectively from U_a and U_b.

Step 2   Select one node included in D_a randomly from *child*(*u*), and let the selected node denote u^*.

Step 3   Select one terminal node included in $st(u^*)$ and located in the deepest place, by a roulette selection relevant to the number of NULL. Let the selected node denote u^{**}.

Step 4   Assign a terminal symbol to *parent*(u^{**}) according to the probability table, and delete u^{**} and its sibling nodes.

Insert is the operator applied when $arg(u) < arg(v)$, which inserts nodes to the node *u* as descendants. The operation of *Insert* is as follows.

Insert (u, v)

Step 1   Select one pair of nodes *u* and *v* that satisfy $arg(u) < arg(v)$ and are located in the same gene locus, respectively from U_a and U_b.

Step 2   Generate $arg(v) - arg(u)$ terminal nodes without assigning symbols, and append to *u* as its child nodes. Let the set of these new nodes be *N*.

Step 3   Substitute $symbl(v)$ for $symbl(u)$.

Step 4   Assign a symbol to the node included in *N*, in accordance with its corresponding probability table.

Step 5   For each node included in *N*, append nodes with symbols assigned according to the probability table as descendants until all leaves become terminal symbols.

3.3.3   Generation Method of Neighborhood Solutions

At the step *k* in the local search from the parent p_1 to the parent p_2 in dMSXF, neighborhood candidates y_i of x_k ($1 \leq i \leq \mu$) are generated by *Replace*, *Delete* and *Insert* between the solution x_k and the target p_2. Each neighborhood solution y_i of x_k is generated as follows: (1) Copy x_k to y_i, (2) Apply *Replace*, *Delete* or *Insert* to y_i until y_i satisfies $d(y_i, x_k) \geq d_{max}$, where d_{max} is selected randomly with the range of $[1, 2 \times d(p_1, p_2)/k_{max} - 1]$ at each step of the local search.

3.3.4   Generation Alternation Model

The generation alternation model we used in this paper is outlined below. This model focuses on a local search performance [8–10].

Generation Alternation Model

Step 1   Generate the initial population composed of N_{pop} of random solutions, individuals, $\{x_1, x_2, \ldots, x_{N_{pop}}\}$.

Step 2   Reset indexes $\{1, 2, \ldots, N_{pop}\}$ to each individual randomly.

Step 3   Select N_{pop} pairs of parents (x_i, x_{i+1}) ($1 \leq i \leq N_{pop}$) where $x_{N_{pop}+1} = x_1$.

Step 4   Apply dMSXF to each pair (x_i, x_{i+1}).
Step 5   For each pair (x_i, x_{i+1}), select the best individual c from offspring $C(x_i, x_{i+1})$
generated by parents (x_i, x_{i+1}) and replace the parent x_i with c.
Step 6   Go to 2 until some terminal criterion is satisfied, e.g., generations and/or
the number of evaluations.

4 Numerical Experiment

Here, we note dMSXF with the conventional PPT and dMSXF with the proposed
UPT as dMSXF+PPT and as dMSXF+UPT, respectively. To show the effectiveness of
incorporation of UPT to dMSXF, we compare dMSXF+UPT to the original dMSXF
and dMSXF+PPT, in symbolic regression problems.

4.1 Problem Domain and Instance

Let V^{NT} and V^T denote the set of nonterminal nodes and the set of terminal nodes, respectively. We employed two functions as below to evaluate the search performance.
Function (I): $f = x^4 - x^3 + x^2 - x$
$V^{NT} = \{+, -, \times, /\}$ and $V^T = \{x, 0.0, 0.1, 0.2, 0.3, 0.4, 0.5, 0.6, 0.7, 0.8, 0.9\}$ are
used to express the function, where x is a variable.
Function (II): $f = x^8 + x^5 + 3\cos x + x$
$V^{NT} = \{+, -, \times, /, \sin, \cos\}$ and the same V^T as Function (I) are used to express
the function.

In these instances, sin and cos are unary functions that have one child node, and
other arithmetic operators are binary functions that have two child nodes. To solve
them, a binary complete tree is required in order to express the PPT.

These functions are estimated from the training set consisting of 21 data points
that are placed at equal interval in the domain $[-1, 1]$. The objective function is
the sum of error at each sample point, and GP minimizes the objective function.
In these regression problems, individuals often include operation exceptions, e.g.,
division by zero. Here, we set the objective function value to ∞, when the individual
includes such an operation. Note that there are several ways to express Function (I)
and Function (II). We assume that the population reaches the optimal solution when
the estimate function becomes completely equivalent to the defined function.

4.2 Experimental Setting

We conducted 100 runs of GP in advance for obtaining the probabilistic trees, and
both UPT and PPT were constructed from top M runs that had reached good solutions

with a high fitness among the 100 runs. In this procedure, dMSXF was employed as the crossover. UPT and PPT were initialized in conjunction with the initialization of the population in the first run of the M runs. During each run, they were updated with upper half of the neighborhood solutions generated in every local searches in terms of the objective function value. UPT and PPT of the previous run were carried over to the next run.

After developing the probabilistic trees, we compared the search performance of dMSXF, dMSXF+PPT and dMSXF+UPT with another 100 runs. In the experiments, the population size was set to 50 and each run was terminated after 200 generations. In three methods, k_{max} was set to 3 and μ was 4 for both instances. These settings were derived from preliminary experiments. In dMSXF, the initial solutions were generated randomly, where every symbols to be assigned to each node were selected with equal probability. In dMSXF+UPT and dMSXF+PPT, initial solutions were generated according to the probability based on the probabilistic tree. An initial tree, a solution of the initial population, was generated with a size in the range of [25, 35] nodes. The depth of PPT was set to 7 which is enough deep to express optimal solutions of Function (I) but not deep enough against those of Function (II).

4.3 Comparing Results

Figure 3 shows the success ratio (%Success) of runs that reach optimal solutions out of 100 runs, in Function (I) and Function (II). In these results, M was set to 20, where M means the number of trials used for constructing probabilistic trees. Figure 4 shows the effect of the number of trials used for constructing probabilistic tree in Function (I). Table 1 compares the depth and the size of the probabilistic trees between dMSXF+PPT and dMSXF+UPT when M was set to 20. In Table 1, the depth of UPT shows the max depth of obtained UPT. The size means the number of nodes in the probabilistic tree.

Fig. 3 Comparison of success ratio

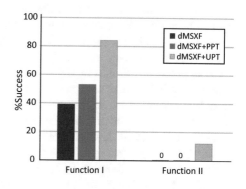

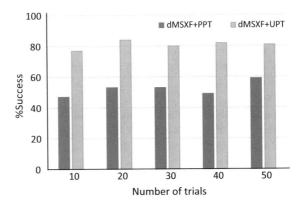

Fig. 4 Effect of the setting of M (Function (I))

Table 1 Comparison of the tree size (M=20)

	Function (I)	Function (II)
PPT	7 (127)	7 (127)
UPT	10 (631)	14 (1288)
		depth (size)

From the result in Function (I) shown in Fig. 3, we can see that both probabilistic trees improve the search performance of dMSXF, and UPT outperforms PPT. In addition, this tendency can be found regardless of the setting of M as shown in Fig. 4.

A remarkable difference of the performance of UPT and PPT is observed in Function (II); dMSXF+UPT can find optimal while dMSXF+PPT cannot. This is because the depth of optimal solutions of Function (II) is more than 8 as shown in Fig. 5, and the depth of PPT is too small to keep good traits for composing optimal solutions. The depth of UPT is more than the required depth for expressing optimal solutions as shown in Table 1, so that some of runs with dMSXF+UPT reached optimal solutions. In another experiment, we confirmed that dMSXF+PPT can find optimal solutions in Function (II) when the depth was set to more than 8.

By using probabilistic trees, optimal solutions are found even though they are not found in previous 100 runs by dMSXF for constructing PPT and UPT. In the local search of dMSXF, there are many individuals that include the exceptions, which degrades the efficiency of the search. The probabilistic trees suppress the appearance of such a lethal trait and enhance the probability of building optimal traits.

UPT can change its depth suitable for the problem through the search. However, the number of node becomes large as shown in Table 1. Many nodes with a high value of NULL were found in UPT. The frequency of appearance of such a node is small, and its probabilistic table might not work well in deciding the symbol because the number of samples is small. Thus, a pruning technique would be effective to suppress the computation cost.

Fig. 5 Example of optimal solution of function (II)

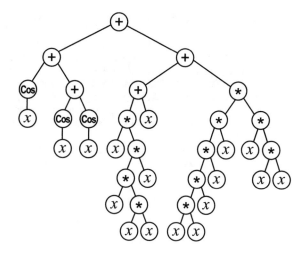

5 Conclusion

In this paper, we proposed union probabilistic tree (UPT) that changes the depth and the breadth suitable for problems, by introducing a union operator between trees. Here we incorporated UPT to dMSXF that is one of local search-based crossovers and evaluated the search performance in symbolic regression problems. Through the numerical experiments, we confirmed that UPT much enhanced the search performance of GP compared to the conventional probabilistic prototype tree (PPT) expressed as a perfect tree. The proposed UPT grows including useless nodes through the search and a pruning technique is also required. This task is left as a future goal.

Acknowledgements This work was supported by JSPS KAKENHI Grant Number 26330290.

References

1. Sałustowicz, R.P., Jurgen Schmidhuber, J.: Probabilistic incremental program evolution. Evol. Comput. **5**(2), 123–141 (1997)
2. Majeed, H., Ryan, C.: A less destructive, context-aware crossover operator for GP. In: Proceedings of the 9th European Conference on Genetic Programming (EuroGP), Lecture Notes in Computer Science, vol. 3905, pp. 36–48 (2006)
3. Hasegawa, Y., Iba, H.: A Bayesian network approach to program generation. Proc. IEEE Trans. Evol. Comput. **12**(6), 750–764 (2008)
4. Beadle, L., Johnson, C.G.: Semantically driven crossover in genetic programming. In: Proceedings of IEEE World Congr. Comput. Intell. (WCCI), vol. 2008, pp. 111–116 (2008)
5. Sato, H., Hasegawa, Y., Bollegala, D., Iba, H.: Probabilistic model building GP with belief propagation. In: Proceedings of WCCI 2012 IEEE World Congress on Computational Intelligence, pp. 2089–2096 (2012)

6. Yanai, K., Iba, H.: Estimation of distribution programming based on Bayesian Network. In: Proceedings of Congress on Evolutionary Computation, vol. 2003, pp. 1618–1625 (2003)
7. Matsumura, K., Hanada, Y., Ono, K.: Probabilistic model-based multistep crossover considering dependency between nodes in tree optimization. In: Studies in Computational Intelligence book series, vol. 721, pp. 187–200 (2017)
8. Ikeda, K., Kobayashi, S.: Deterministic multi-step crossover fusion: a handy crossover for GAs. In: Proceeding of Parallel Problem Solving from Nature VII, pp. 162–171 (2002)
9. Hanada, Y., Hiroyasu, T., Miki, M.: Genetic multi-step search in interpolation and extrapolation domain. In: Proceedings of the Genetic and Evolutionary Computation Conference, vol. 2007, pp. 1242–1249 (2007)
10. Nagata, Y.: New EAX crossover for large TSP instances. In: Proceedings of the Parallel Problem Solving from Nature (PPSN) IX, pp. 372–381 (2006)

Infrastructure in Assessing Disaster-Relief Agents in the RoboCupRescue Simulation

Shunki Takami, Masaki Onishi, Itsuki Noda, Kazunori Iwata, Nobuhiro Ito, Takeshi Uchitane and Yohsuke Murase

Abstract The RoboCupRescue Simulation project has been implemented as one of the responses to recent large-scale natural disasters. In particular, the project provides a platform for assessing disaster-relief agents and simulations. However, its research evolution is limited because all agents' programs must be developed by each researcher and the experimental operations are complex. To address these problems, we propose a combination of an agent development framework and experiment management software in this study as infrastructures in assessing disaster-relief agents in the RoboCupRescue Simulation. We have provided those elements separately; however, it becomes possible to easily carry out experiments that have flexible configuration by combining two elements. In the evaluation, a combinatorial experiment

S. Takami (✉)
University of Tsukuba, 1-1-1 Tennodai, Tsukuba, Ibaraki 305-8577, Japan
e-mail: s-takami@aist.go.jp

M. Onishi · I. Noda
National Institute of Advanced Industrial Science and Technology (AIST),
2-3-26 Aomi, Koto-ku Tokyo, 135-0064, Japan
e-mail: onishi@ni.aist.go.jp

I. Noda
e-mail: i.noda@aist.go.jp

K. Iwata
Aichi University, 4-60-6 Hiraike-cho, Nakamura-ku, Nagoya, Aichi 453-8777, Japan
e-mail: kazunori@vega.aichi-u.ac.jp

N. Ito · T. Uchitane
Aichi Institute of Technology, 1247 Yachigusa,
Yakusa-cho, Toyota Aichi, 470-0392, Japan
e-mail: n-ito@aitech.ac.jp

T. Uchitane
e-mail: takeshi.uchitane@aitech.ac.jp

Y. Murase
RIKEN Center for Computational Science, 7-1-26 Minatojima-minami-machi,
Chuo-ku, Kobe, Hyogo 650-0047, Japan
e-mail: yohsuke.murase@riken.jp

© Springer Nature Switzerland AG 2020
R. Lee (ed.), *Computational Science/Intelligence and Applied Informatics*,
Studies in Computational Intelligence 848,
https://doi.org/10.1007/978-3-030-25225-0_10

133

as a case study confirms the effectiveness of the environment and shows that the environment can contribute to future disaster response research that utilizes a multi-agent simulation.

1   Introduction

The international RoboCup community has continued the RoboCupRescue Simulation (RRS) project to confront large-scale natural disasters since 2001 [9, 17]. In particular, the RRS Agent Competition is an annual program for assessing disaster-relief agents and simulations. This project aims to provide in-depth the knowledge on disaster-relief and to share the results of studies with society [21].

However, researchers should implement many types of algorithms to disaster-relief agents, such as that for path planning, information sharing, and resource allocation, to address these disaster-relief problems targeted by the RRS. Therefore, researchers exchange their gathered technical information through the competition. Effective algorithms are shared via agent program codes and the documents that describe the design policy of the agent with us. However, these codes are difficult to be included on each researcher' s agent because a structure of program codes differ from one researcher to another. Moreover, the effectiveness of these algorithms must be investigated on many disaster scenarios in various conditions, and these investigations must involve a large number of simulations.

Therefore, the combination of agent development framework, the RRS-ADF and experiment management software, RRS-OACIS is proposed as an infrastructure to assess disaster-relief agents in the RRS. Through experiments, we evaluate whether the infrastructure is useful to assessing disaster-relief agents. This framework integrates the structure of agent codes within the RRS to modularize each algorithm [20]. Meanwhile, this management software supports the implementation of experiments and the setup of simulation environment using multiple computer clusters [19]. We have provided those elements separately. Several minor adjustments are made to the framework and the management software to allow them to work together. In the evaluation, a combinatorial experiment as a case study confirms the effectiveness of the environment and shows that this environment can contribute future disaster response research that utilizes a multi-agent simulation.

2   Multi-agent Simulation and RoboCupRescue

2.1   Multi-agent Simulation

Majority of studies on artificial intelligence adopt a method that investigate the effectiveness of each proposed method on a simplified virtual space. In contrast, a simulation that introduces a multi-agent system reproduces an autonomous agent with many algorithms in a highly realistic virtual world. Thus, the implementation of this simulation tends to be complex, the experiment is large-scale, and the research difficulty is remarkable.

2.2 RoboCupRescue Simulation

The RRS is a research platform that simulates disaster situations and disaster-relief activities on a computer. Figure 1 shows the activities of agents in the RRS. In these activities, researchers control six types of agents, namely AmbulanceTeam, FireBrigade, PoliceForce, and the headquarters of each unit. In addition, there are agents to simulate disaster situations, namely Civilian.

– **AmbulanceTeam and AmbulanceCentre**
  These agents rescue other agents that cannot move on their own.
– **FireBrigade and FireStation**
  These agents extinguish fires in buildings.
– **PoliceForce and PoliceOffice**
  These agents clear road blockages.
– **Civilian**
  In the competition, these agents move automatically to evacuation centers.

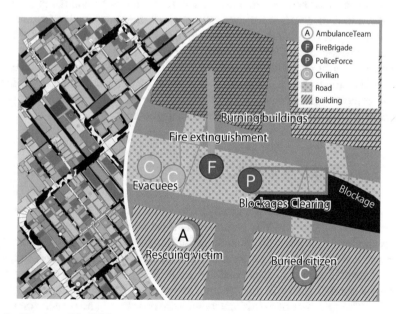

Fig. 1 Overview of the RRS

Through this RRS, investigating the applications of artificial intelligence and information science to natural disaster rescue problems is possible. Hence, researchers have been evaluating algorithms for route searching, information sharing, and task allocation during a disaster. In the RRS project, five tasks are supported, namely group formation, path planning, search, multi-task allocation, and communication [18]. The competition involving agent program codes are facilitated annually for technical exchange purposes [2].

2.3   Agent Development in the RRS

The disaster-relief problems handled by the RRS are complex because damage situations, such as fire, building collapse, and availability of radio communication, periodically changes in affected areas. These changes are addressed via disaster-relief strategies of teams of disaster-relief robots, which vary depending on disaster situations. To implement a disaster-relief strategy, it is necessary to prepare all the algorithms for various tasks such as route searching, information sharing, and resource allocation in the disaster environment. Moreover, it is necessary to implement control by angles and coordinates for activities such as blockages being cleared by the PoliceForce. Those controls will be a hurdle in the implementation of agents if a researcher wants to study the strategy.

To promote the research involving the RRS, the structure of a complex disaster-relief problem must be clarified and be subdivided before solving the problem. In the RRS, although, a researcher can freely implement a source code in considering that the agent can communicate with the simulator's kernel, the implemented codes should be integrated with the other codes. To realize that a researcher can share the implementation of his or her respective research fields and another researcher can reuse it, the structure of a complex disaster-relief problem should be divided into several components associated with the research fields. Thus, the components as implementations in each field should be reusable modules.

2.4   Experiments in the RRS

To develop and evaluate the RRS agents, experiments on multiple disaster areas must be performed while considering various conditions such as locations of fires, rate at which buildings collapse, and communication situations. Moreover, the parameters of an agent may also be set for each of these situations. Hence, a large number of simulations are required to obtain findings. In the RRS, execution of each simulation consumes approximately 20 min. Agents' calculation requires sufficient memory capacity. The processes of agents' calculation are distributed into several computers and executed in the distributed computers.

Generally, researchers have manually executed these experimental processes. However, the number of experiments should be remarkably increased to promote research, and this manual experiment management becomes difficult.

2.5 Related Work

2.5.1 OpenRTM-Aist

It is a framework for robot development developed by the National Institute of Advanced Industrial Science and Technology [12], and its implementation is in accordance with the RT-Middleware standard [3]. This platform divides robot elements, such as actuator control, sensor input, and algorithms, necessary for behavior control into single components, known as RT-Components (RTCs). RT-Middleware also develops a robot by combining all such the RTCs. Hence, subdividing the elements that are necessary for controlling the robot is possible. Considering that each component can be exchanged as a module and existing modules can be included, the challenges encountered by developers when developing and improving robots can be possibly reduced.

Given that RT-Middleware is applied mainly to real robots, developing robots that are controlled in real time is appropriate. However, utilizing the existing code and knowledge is difficult when adopting RT-Middleware because the RRS agents are programmed mainly with a sequential structure.

2.5.2 OACIS

It is a simulation-execution management framework developed by the Discrete Event Simulation Research Team of the RIKEN Center for Computational Science [11]. OACIS has the function of managing jobs. Particular, it has a job-management function that specializes in simulation execution and a management function of the experimental results. Moreover, OACIS supports a large number of simulations and analyzes various conditions by automatically managing the experimental parameters and results.

However, complex operations are required to execute the RRS simulations because OACIS is a general-purpose system for various types of simulation software. The construction of simulation scripts, agent programs, and disaster scenario files must be managed outside OACIS.

3   System and Experiments

3.1   *Research Objective*

In this study, we evaluate the combination of agent development framework and experiment management software to confirm whether it is useful as an infrastructure to assess the disaster-relief agents in the RRS. The agent development framework unifies the structure of agent codes within the RRS to modularize each algorithm, whereas this management software supports the implementation of experiments and the setup of simulation environment using multiple computer clusters. It is evaluated through experiments whether the infrastructure is useful to assessing disaster-relief agents.

   In this section, we describe the framework and the software, and finally describe the experiments.

3.2   *Agent Development Framework*

3.2.1   Design Policy of the Agent Development Framework

This framework modularizes part of the program code to clarify and address complex problems. An agent development framework is necessary to ensure that rescuing the program code becomes easy and to reduce the responsibility of researchers. A framework design method based on OpenRTM-aist would also be feasible. However, the RRS agents are programmed in a sequential structure. A possibility that current researchers will not use our framework if the programming structure is substantially modified is also a concern. Therefore, we propose and design a unique agent development framework in which utilizing the existing program code and knowledge becomes easy. We decided to call this Agent Development Framework for RoboCupRescue Simulation (RRS-ADF). This RRS-ADF provides a common architecture for agents, program code modularization, aggregation of information acquisition interfaces necessary for agent decision-making, and unified inter-agent communication protocol.

3.2.2   Common Agent Architecture

The RRS-ADF reduces the differences in combinations of components by each developer and ensure their re-usability by defining the overall behavior of an agent as a common architecture. This approach allows developers to implement modules based on common architectures when developing agent programs.

   Figure 2 shows the conditions before and after introducing the RRS-ADF. The left-hand portion of the figure presents the examples of existing agent structure,

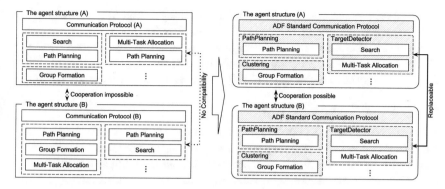

Fig. 2 Schematic of before (left) and after (right) introducing the RRS-ADF

whereas the right-hand part introduces the common architecture. The portability of the existing program code is low because each researcher independently develops an agent program according to his or her individual research agendas. Therefore, we have commonized the structure of the agent program (indicated by the shaded part of the figure) to ensure that the program code can be re-used easily. This agent program structure integrates the inter-agent communication protocol and enables the communication with agents developed by other researchers.

Similar to RT-Middleware discussed in Sect. 2.5.1, modularization of program codes is introduced to minimize the problem of researchers and to ensure the possibility of re-using the program codes utilized in the agent development.

Algorithm modularization

The RRS-ADF is divided to the extent possible on the basis of the five tasks presented in the RRS project. The framework provides PathPlanning, Clustering, and TargetDetector modules to each rescue-agent unit. Figure 3 illustrates the relationship between the RRS tasks and the RRS-ADF modules. The left-hand portion of the figure contains five tasks, whereas the right-hand part involves the framework

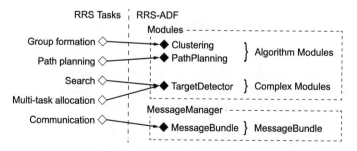

Fig. 3 Relationships between the RRS tasks (left) and the RRS-ADF modules (right)

modules. PathPlanning and Clustering modules belong to Algorithm Modules group and TargetDetector module belongs to Complex Modules group.

We classify the algorithms for solving complex and simple problems as Complex Modules and Algorithm Modules, respectively; thus, directionality of each module is clarified. Complex Modules can probably divide the internal structure in the future.

Modularization of the control program

As described in Sect. 2.3, controlling an agent is necessary by specifying such properties as its coordinates and angles. A low-level agent control is modularized as a control program. The framework provides ActionFireFighting, ActionExtClear, ActionTransport, and ActionExtMove modules as low-level agent control modules. Owing to the limitation of space, the detailed function of these module was omitted. By separating macro algorithms (e.g., decision making) and micro algorithms (e.g., control) using the coordinates and angles, we minimize the burden of researchers who are interested in investigating a single algorithm, such as decision making.

3.2.3 Aggregation of Information-Acquisition Interfaces

The framework aggregates the interfaces that acquire the information necessary for agent decision making provided by the kernel. It clarifies the acquired data.

3.2.4 Collective Management of Parameters

The framework collectively manages the specifications of the modules to be used and the eigenvalues in the algorithm to be changed during experiments. Thus, managing various parameters becomes easy. The parameters in the algorithms can be obtained from dedicated components, and each value can be input by using JavaScript Object Notation [6]-formatted text as an argument at the agent startup. By utilizing these interfaces, the agents can cooperate with the experiment management software, which will be discussed later.

3.2.5 Unifying Inter-agent Communication Protocols

The inter-agent communication protocols in the RRS are currently not integrated. This situation reinforces the dependency among components; however, modularization of the algorithm becomes challenging. Therefore, we define a common inter-agent communication protocol based on the RRS inter-agent communication protocols proposed by Ota et al. [14] and Obashi et al. [13].

We define the messages communicated in this protocol as members of either an information-sharing or command family. In the information-sharing family, information related to agents, roads, and buildings is shared. Meanwhile, in the command family, the commands for relief, fire extinguishing, blockage clearing, and searching are implemented.

3.3 Software Used for Experiment Management

3.3.1 Design Policy of the Experiment Management Software

Figure 4 illustrates the experimental processes of the RRS and the parts that the software automates. This software was developed on the basis of OACIS described in Sect. 2.5.2, referred to as RRS-OACIS in this paper. The RRS requires many manual operations, as indicated by the black dots in the figure. The operations specify various agents' parameters, control computer clusters, and collect numerous simulation results. RRS-OACIS can automate the operations, as denoted by the shaded area in the figure. The RRS-ADF exploitation is simple owing to the software implemented for the RRS. Moreover, repetition of experiments becomes easy by automating the operations. The RRS-OACIS implementation was thoroughly designed to ensure maintainability. It does not modify the OACIS code because all control runs on application programming interfaces. Therefore, this section describes the points added in RRS-OACIS in comparison with OACIS.

3.3.2 Agent Management

Although OACIS can manage simulators and experimental parameters, it cannot manage agent program files. In contrast, RRS-OACIS has a function that can regulate these files.

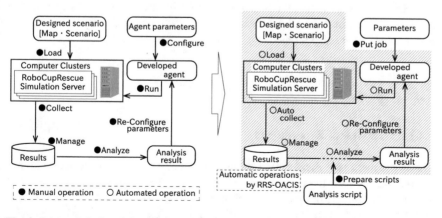

Fig. 4 Schematic before (left) and after (right) introducing RRS-OACIS

3.3.3 Map and Scenario Management

OACIS cannot handle files related to maps and disaster scenarios. In contrast, RRS-OACIS implements a function that can manage these files.

3.3.4 Computer Cluster Management

In the RRS, one simulation is executed in a computer cluster. Although OACIS can activate most simulations within a computer cluster, it cannot directly run the RRS. In contrast, RRS-OACIS has a function that bridges the activation processes between OACIS and the RRS.

3.3.5 Simulation Script

OACIS cannot handle complex processes combined with multiple operations using multiple computers in a simulation. Therefore, a script that describes a series of operations in a simulation should be prepared. This script initially loads agent programs, map files, and scenario files, and, then, it is connected to each computer in the cluster and executes the simulation. The script establishes an experiment as a single job in OACIS that OACIS can manage. Moreover, RRS-OACIS works with the RRS-ADF through this script.

3.3.6 Simulator Management

OACIS considers a simulation script and a set of parameters as part of a simulator. In other words, the script and parameters are embedded into a simulator. The parameters differ depending on the experimental purpose, modules, and algorithm parameters. RRS-OACIS has the function that can install plug-ins according to the experimental purpose. A plug-in can be independently developed, and general-purpose experimental plug-ins are also provided.

3.4 Experiments

3.4.1 Overview of Experiments

The experiment was divided into three phases because there are a large number of combinations for the experiments are possible. The target teams were selected in the first phase, partial module combinations were selected in the second phase, and the best module combination was determined in the final phase. We eventually experimented with two (Eindhoven and Paris) scenarios using the selected target teams. The experiment was performed on the teams who participated in RoboCup 2017 [16]. The participating teams that year developed an agent using the RRS-ADF according to our proposal [7].

3.4.2   Selection the Target Teams

All (twelve) the teams of RoboCup 2017 ran on RRS-OACIS. Each team was experimented on five randomly chosen scenarios from the scenarios used in RoboCup 2017. The experiments of the same scenario were performed fifteen times. These median values were defined as the team evaluation values and ranked according to the official rule. The target teams are the top five performing teams in the experiment as follows: MRL [4], Aura [8], RoboAKUT [1], SEU-UniRobot [15], and CSU_Yunlu [22].

3.4.3   Selection Partial Module Combinations

The modules of each target team are used with other modules that are necessary in experimenting the combination pattern. These modules are collectively called the base team in this paper. The combinations use the patterns, P_a, P_b, and P_c, as defined as follows:

P_a: BuildingDetector, RoadDetector, and HumanDetector;
P_b: Search for FireBrigade, Search for PoliceForce, and Search for Ambulance;
P_c: ActionFireFighting, ActionExtClear, ActionTransport, ActionExtMove, and corresponding PathPlanning modules.

The experiments of the same parameter were performed nine times. The median values were defined as team evaluation values and ranked in descending order. The combinations of the selected partial modules are omitted due to space limitations.

3.4.4   Determination of the Best Module Combination

The best combination was determined by performing a combinatorial experiment of the partial module combinations, which are selected in the second phase. Meanwhile, the experiments on the same parameter were performed nine times. Similar with the second phase, the median values were defined as team evaluation values and ranked in descending order.

4   Results and Discussion

We only mention the final combination result in this study. Tables 1 and 2 show the results of a series of experiments, that is, the best combination of algorithm modules for each scenario. The best combination of algorithm modules is the designed agent in the experiment. Table 3 presents the score comparison of the designed agent with

Table 1 All the adopted modules for the Eindhoven scenario

Module type	Team
BuildingDetector	MRL
RoadDetector	MRL
HumanDetector	Aura
Search (FireBrigade)	CSU_Yunlu
Search (PoliceForce)	MRL
Search (AmbulanceTeam)	Aura
ActionFireFighting	MRL
ActionExtClear	MRL
ActionTransport	RoboAKUT
ActionExtMove	Aura

Table 2 All the adopted modules for the Paris scenario

Module type	Team
BuildingDetector	Aura
RoadDetector	Aura
HumanDetector	MRL
Search (FireBrigade)	MRL
Search (PoliceForce)	MRL
Search (AmbulanceTeam)	Aura
ActionFireFighting	Aura
ActionExtClear	Aura
ActionTransport	RoboAKUT
ActionExtMove	Aura

Table 3 Scores of the base teams and designed agent for each scenario

\ Team Scenario	MRL	Aura	Robo AKUT	SEU-UniRobot	CSU_ Yunlu	Designed agent
Eindhoven	96.05	115.44	93.18	88.36	94.57	116.40
Paris	41.03	42.38	33.19	19.62	18.46	61.06

that of several original agents of RoboCup 2017 for each scenario. The higher the score in the RRS, the better. These results confirm the effectiveness of the designed agent.

We could develop agents with excellent performance through combinatorial module experiments developed by various researchers. In the Eindhoven scenario, there is no significant difference between the agent developed by the experiment and the original agent (Aura) because Aura's strategy fitted to the scenario. That a strategy has a degree of fitness to a scenario has appeared as the difference of the selected modules between Eindhoven and Paris. After all, it consists of a combination of modules developed by multiple researchers.

The experiments were automatically facilitated and completed in approximately 50 days without problem. These results demonstrate that our environment is an effective approach for developing and evaluating the algorithms of multi-agent systems. To evaluate disaster response ideas, an evaluation framework is necessary to evaluate its effectiveness. Hence, our infrastructure is assumed to contribute to the multi-agent and rescue engineering research community in this respect.

Based on these results, we confirmed the effectiveness of the environment and show that this environment can contribute future disaster response research that utilizes a multi-agent simulation. However, it is necessary to be able to evaluate the influence rate of each module in order to analyze the relationship of modules in more detail. We think this issue is the task we should solve next. Moreover, it is necessary to introduce methods such as a genetic algorithm in order to reduce the calculation time because the modules selection method in this paper is needed too long time.

5 Conclusion

In this study, we have proposed the combination of the RRS-ADF which is an agent development framework, and RRS-OACIS, which is an experiment management software as an infrastructure to assess the disaster-relief agents in the RRS. The study aims to promote research in disaster response that utilizes a multi-agent simulation. The case study of a set of the combinatorial experiments demonstrates the effectiveness of the environment.

The proposed the RRS-ADF and RRS-OACIS are used by researchers in various countries as the RRS community official tool since 2017. Specifically, the use of the RRS-ADF is a prerequisite in the participation of the annual competition. In addition, the implementation examples in which MATLAB [10] of MathWorks and the RRS-ADF are linked have been reported from other researchers [5]. RRS-OACIS is also utilized for competition management.

Our aim is to provide the research results obtained in the RRS project available to society by clarifying disaster-relief problems and proposing algorithms that are applicable for an effective disaster-relief. Therefore, our environment can contribute to the multi-agent systems and rescue engineering research communities. We will also continuously enhance this infrastructure while rendering it available to the RRS community.

Acknowledgements This work was partially supported by MEXT Post-K project "Studies of multi-level spatiotemporal simulation of socioeconomic phenomena". This work was supported by JSPS KAKENHI Grant Number JP16K00310 and JP17K00317. The authors would like to thank Enago (www.enago.jp) for the English language review.

References

1. Akın, H.L., Aşık, O.: Robocup 2017 rescue simulation league team description RoboAKUT (turkey). http://roborescue.sf.net/data/2017-RSL/agent/rsl17-tdp-agent-RoboAKUT.pdf (2017). Accessed 10 Feb 2019
2. Akin, H.L., Ito, N., Jacoff, A., Kleiner, A., Pellenz, J., Visser, A.: Robocup rescue robot and simulation leagues. AI Magazine **34**(1), 78–86 (2013). URL http://dblp.uni-trier.de/db/journals/aim/aim34.html#AkinIJKPV13
3. Ando, N., Suehiro, T., Kitagaki, K., Kotoku, T., Yoon, W.: Rt-middleware: distributed component middleware for rt (robot technology). In: 2005 IEEE/RSJ International Conference on Intelligent Robots and Systems, pp. 3933–3938 (2005). https://doi.org/10.1109/IROS.2005.1545521
4. Ardestani, P., Taherian, M., MohammadAliZadeh, P., Nikoo, E.J.: Robocup rescue 2017- agent simulation League team description paper MRL (iran). http://roborescue.sf.net/data/2017-RSL/agent/rsl17-tdp-agent-MRL.pdf (2017). Accessed 10 Feb 2019
5. Castro, S.: Machine learning to the rescue!. https://blogs.mathworks.com/racing-lounge/2018/07/04/machine-learning-to-the-rescue/ (2018). Accessed 10 Feb 2019
6. ECMA International: Standard ecma-404 2st edition (the JSON data interchange syntax). https://www.ecma-international.org/publications/files/ECMA-ST/ECMA-404.pdf (2017). Accessed 10 Feb 2019
7. Faraji, F., Nardin, L.G., Modaresi, A., Helal, D., Iwata, K., Ito, N.: Robocup rescue simulation league agent 2017 competition rules and setup. http://roborescue.sourceforge.net/web/2018/downloads/rules2017.pdf. Accessed 10 Feb 2019
8. Ghahramanpour, M., Absalan, A., Kandeh, A.: Robocup 2017 rescue simulation league team description aura (iran). http://roborescue.sf.net/data/2017-RSL/agent/rsl17-tdp-agent-Aura.pdf (2017). Accessed 10 Feb 2019
9. Kitano, H., Tadokoro, S.: Robocup rescue: a grand challenge for multiagent and intelligent systems. AI Mag. **22**(1), 39 (2001)
10. MathWorks, Inc.: Matlab overview. https://www.mathworks.com/products/matlab.html. Accessed 10 Feb 2019
11. Murase, Y., Uchitane, T., Ito, N.: A tool for parameter-space explorations. Phys. Procedia **57**, 73–76 (2014). https://doi.org/10.1016/j.phpro.2014.08.134
12. National Institute of Advanced Industrial Science and Technology: Openrtm-aist. http://www.openrtm.org/ (2019). Accessed 10 Feb 2019
13. Obashi, D., Hayashi, T., Iwata, K., Ito, N.: An implementation of communication library among heterogenous agents naito-rescue 2013(japan). In: RoboCup 2013 Eindhoven (2013)
14. Ohta, T., Toriumi, F.: Robocuprescue2011 rescue simulation league team description. In: RoboCup 2011 Istanbul (2011)
15. Qian, C.: Robocup 2017 rescue simulation league team description SEU-UniRobot (P.R.China).http://roborescue.sf.net/data/2017-RSL/agent/rsl17-tdp-agent-SEU-UniRobot.pdf (2017). Accessed 10 Feb 2019
16. RoboCupRescue Simulation League 2017 Committee: Robocup 2017 agent simulation results. https://rescuesim.robocup.org/2017-nagoya-agent-results/ (2017). Accessed 10 Feb 2019
17. RoboCupRescue Simulation League Committee: Robocuprescue simulation. http://rescuesim.robocup.org/. Accessed 10 Feb 2019
18. Skinner, C., Ramchurn, S.: The robocup rescue simulation platform. In: Proceedings of the 9th International Conference on Autonomous Agents and Multiagent Systems: Volume 1 - Volume 1, AAMAS '10, pp. 1647–1648. International Foundation for Autonomous Agents and Multiagent Systems (2010)
19. Takami, S., Onishi, M., Iwata, K., Ito, N., Murase, Y., Uchitane, T.: An environment for combinatorial experiments in a multi-agent simulation for disaster response. In: PRIMA 2018: Principles and Practice of Multi-Agent Systems - 21st International Conference, Tokyo, Japan, 29 Oct 29–2 Nov 2018, Proceedings, pp. 646–654 (2018). https://doi.org/10.1007/978-3-030-03098-8_52

20. Takami, S., Takayanagi, K., Jaishy, S., Ito, N., Iwata, K.: Agent-development framework based on modular structure to research disaster-relief activities. IJSI **6**(4), 1–15 (2018). https://doi.org/10.4018/IJSI.2018100101

21. Visser, A., Ito, N., Kleiner, A.: Robocup rescue simulation innovation strategy. In: RoboCup 2014: Robot World Cup XVIII, pp. 661–672 (2015). https://doi.org/10.1007/978-3-319-18615-3_54

22. Zhang, P., Kang, T., Jiang, F., Peng, J., Zhang, X.: Robocup 2017 rescue simulation league team description CSU_yunlu (china). http://roborescue.sf.net/data/2017-RSL/agent/rsl17-tdp-agent-CSU_Yunlu.pdf (2017). Accessed 10 Feb 2019

OOCQM: Object Oriented Code Quality Meter

Asma Shaheen, Usman Qamar, Aiman Nazir, Raheela Bibi, Munazza Ansar and Iqra Zafar

Abstract Source code is core of Software engineering. Source code of good quality can be maintained and upgraded easily. Source code quality can be broken down in factors like maintainability, reusability, change proneness, testability and stability. These quality factors are high level representation of code quality and cannot be measured directly. Research in field of source code quality has proposed enormous amount of code metrics that measure quality in different aspects like complexity, size, coupling, cohesion and abstractness etc. This study proposes a framework named Object Oriented Code Quality Meter (OOCQM) for measuring source code quality of object-oriented code using low level code metrics and high-level quality factors. Code metrics has a proven relationship with quality factors. This relationship is used to calculate a numerical value for quality factors based on metric values. It is observed that all selected metrics has negative correlation with mapped quality factor. Quality factors scores are aggregated and used to depict quality of code in numerical form. A PHP based tool is developed to validate the results. Framework results are compared with Maintainability Index (MI) which is popular quality measure in terms of code maintainability. OOCQM measures code quality correctly as quality

A. Shaheen (✉)
Department of Computer Engineering, NUST, College of E&ME,
Mirpur AJK, Pakistan
e-mail: asma.shaheen16@ce.ceme.edu.pk

U. Qamar · A. Nazir · R. Bibi · M. Ansar · I. Zafar
Department of Computer Engineering, NUST, College of E&ME,
Rawalpindi, Pakistan
e-mail: usmanq@ceme.nust.edu.pk

A. Nazir
e-mail: aiman.nazir16@ce.ceme.edu.pk

R. Bibi
e-mail: raheela.bibi16@ce.ceme.edu.pk

M. Ansar
e-mail: munazza.ansar16@ce.ceme.edu.pk

I. Zafar
e-mail: iqra.zafar16@ce.ceme.edu.pk

© Springer Nature Switzerland AG 2020
R. Lee (ed.), *Computational Science/Intelligence and Applied Informatics*,
Studies in Computational Intelligence 848,
https://doi.org/10.1007/978-3-030-25225-0_11

results are correct according to MI. This framework provides more detail at individual quality factors level. OOCQM is compared with few other tools developed for quality measurement. Comparison shows that this tool supports more quality factors analysis than other tools.

Keywords OO code quality · Code metrics · Software quality factors · Source code quality calculation

1 Introduction

Software development is quiet mature industry which is flourishing rapidly in many directions. Software is core of modern technology advancements. New technologies like Artificial Intelligence, Machine Learning, and Data Mining etc. are dependent on software. This time is called digital era as digitization is involved in every field of life. Need of digitization has increased use of software exponentially. More and more software are developed by large to small software companies and even individual programmers to cope with the demand of software development. There are many types of software like system software, application software, embedded software, web based and desktop software etc. Software can also be classified based on language in which they are developed.

Software development is a complex and dynamic process which requires a team effort with diverse skills as software development starts from a problem statement and reaches to a state of full fledge deliverable product. Development process includes requirement engineering (requirement gathering and specification), analysis, design, coding, testing and deployment. Quality of a software is a key consideration at every stage of software development. A compromise on quality not only reduces trust on software and on company that delivered software. It is complex to enhance and maintain a low-quality software.

Measuring quality of a software is challenging due to several reasons and few of these reasons are:

- Diversity in software types
- Lots of programming languages
- Lack of standards for quality measures
- Software industry focuses mainly on external quality
- No standard tools available for measuring software quality.

Software quality factors are used as base for measuring quality. A quality factor is a set of non-functional requirements or quality attributes that are well defined and can be grouped in a set.

McCall derived 11 quality factors related to software products. These quality factors can be divided in three groups, factors related to product revision, operation and transition. Correctness, usability, reliability, efficiency and integrity fall in group of factors that are related to software operation. Software transition quality factors are portability, re usability and interoperability. Maintainability, flexibility and testability are software revision related quality factors [1].

Another interesting thing about software quality is that is a multifaceted concept. Software quality can be viewed from five perspectives. These five views include transcendental, user, manufacturing, product and user-based view. Measurement of quality depends on definition of quality from selected perspective of quality [2].

Different quality models have been proposed by researchers working in field of software quality to describe relation between different software characteristics. Models are based on quality factors. Popular models of quality are McCall's quality model and ISO 9126. Software quality models can be divided in types like quality definition models, quality assessment models and quality prediction models. All three types of quality models are criticized for not being widely applicable because requirements for their application are not briefly described [3].

Software quality metrics are used to measure quality of software. A software quality metric is measurable criteria of a software that can indicate goodness or badness of a software. Previous research in field of software code quality provides lot of metrics for measuring different aspects or factors of software. Metric is a lower level quality measure.

There are lot of programming languages and measuring quality of code is somewhat language dependent especially code parsing part is different for every language due to differences in syntax and semantic rules. PHP is an open source web development language that is been used since last thirty years. PHP was developed in 1994 by Rasmus Lerdorf. It was developed in C. PHP has evolved a lot and has become a mature and popular web development language that is easy to learn and efficient in accomplishing web application requirements.

There are more than 300 code quality measurement metrics proposed by researchers during different time periods. Most of the metrics are proposed and implemented on Object Oriented Programming code. There are code metrics for Procedural Programming (PP), Aspect Oriented Programming (AOP) and Feature Oriented Programming (FOP) but those metrics are quite less in number and quite low in popularity as compared to OOP code metrics [4].

The tool developed according to this research is a web-based application that can measure code quality of any PHP project and can present quality results in way that can help developers/managers to take better decisions about project.

Rest of the study has following sequence, Sect. 2 provides the detail description of previous research work done in field of software code quality including quality metrics, quality frameworks and tools and their significance. Section 3 describes the methodology used to design the framework for measuring code quality. Section 4 focuses on validation of proposed framework. Section 5 summarizes the whole research work and also describes the future research possibilities for this study.

2 Related Work

Software code quality metric is a measure that defines any aspect or characteristic of software behavior. Code quality metrics have proven correlation with software characteristics. Metrics provide information to developers about internal characteristics of software that can affect software quality. Many metrics are also beneficial for managers and end users [5].

This section narrates code quality metrics, code quality analysis frameworks and tools used for code quality measurement.

2.1 Software Code Quality Metrics

Following popular metrics are picked from previous research.

2.1.1 SLOC Metric

This metric is considered most basic and simple metric that can be helpful in analyzing code quality. This metric is also used in calculating some other metrics. For example, SLOC can be used for predicting defect density as it has negative correlation with code size [6].

2.1.2 Halstead's Metrics

Halstead's metrics suit is one of oldest metrics suite used to analyze complexity of a software program. According to Halstead a software program is collection of tokens and these tokens fall in two categories operators and operands. Metrics proposed by Halstead are based on counting these two types of tokens. Halstead used operators and proposed ten metrics. Some Halstead's metrics can be used as quality indicators are for example difficulty (D), Effort (E) and Program Time (PT) [7].

2.1.3 Cyclomatic Complexity

Complexity of a software is also used as quality indicator. McCabe used graph theory to represent complexity of source code. According to graph theory code size does not affects its complexity. It means adding new functional statements or removing some functional statements will not change complexity of program. Complexity only depends on decision statements [8].

2.1.4 MOOSE/CK Metrics

This metric suite was proposed by Chidamber and Kemerer in 1991. It was first effort for analyzing object oriented code specifically. There are six metrics in CK metrics suit. Metrics proposed in this suite are most popular in analysis of quality of object-oriented code.

Weighted Methods Per Class (WMC)

CK metric suite provided a new metric for calculating Cyclometic complexity of a class. WMC can be calculated by adding up complexity values of all local methods of a class. WMC is measure of complexity of an object and its value can be used to assess how much time and effort is required to develop and maintain this object. Larger value of WMC indicates that methods in class are not generic and are more related to specific application and this class has low reusability.

Depth of Inheritance Tree (DIT)

DIT is the level of a class in hierarchy of inheritance. Root class has DIT value of 0. The higher DIT means more functionality the class has as it inherits all functionality and properties of super classes. Higher DIT indicates difficulties in maintenance of class. Higher DIT is also symbol of high complexity. Another concern with high DIT is violation of encapsulation as child class can access properties of parent classes.

Number of Children (NOC)

NOC indicates how many classes are directly inherited by a certain class. High NOC means more importance of class in application architecture. Generally speaking, depth is better in hierarchy instead of breath. It means NOC value should not be high especially when class is lower in inheritance hierarchy. If a class has more children then it requires more testing of all methods in class as it has to serve more scenarios.

Coupling Between Objects (CBO)

Coupling refers to access of other class's methods or instances other than inheritance. It refers to access of one class's methods or objects by another class's functions. High CBO means design is not modular and rule of re usability via inheritance is violated. High CBO also make application difficult to test and modify, it also decreases re usability. CBO value indicates how easy or difficult is to maintain and test that class. CBO is measure of communication between objects.

Response for Class (RFC)

RFC denotes to count of numbers of elements that can provide response to an object of class. This set of response elements is collection of all local methods of class plus all methods that are called by local methods of class. Lower RFC values are ideal in terms of low complexity, less testing, easy debugging and easy to understand the code. This metric is related to object attributes and object communication.

Lack of Cohesion in Methods (LCOM)

Cohesion means how well methods of a class are using local variables/properties of that class. LCOM is calculated by counting disjoint sets of methods where one set is made by combining all methods that share a common variable. High LCOM indicates that class should be divided to more classes. This metric is related to attributes of objects [9].

2.1.5   MOOD Metrics

This metrics set also consists of 6 metrics. Metrics in this suit are presented to measure quality of code in terms of OOP features. These metrics measure polymorphism, coupling, encapsulation and inheritance. This suit includes Method Hiding Factor (MHF), Attribute Hiding Factor (AHF), Method Inheritance Factor (MIF), Attribute Inheritance Factor (AIF), Coupling Factor (CF) and Polymorphism Factor (PF) metrics [10].

2.1.6   QMOOD

QMOOD model was proposed for analyzing high level attributes related to quality like re usability, complexity and flexibility. To measure these high-level quality attributes design properties like modularity, cohesion, coupling and encapsulation are used. This model is divided in four levels. First level consists of design quality attributes. Second level contains design properties of object-oriented design. Object oriented design metrics fall in level three and design components of object-oriented design are placed in level four. Authors identified Re usability, Flexibility, Understandability, Functionality, Extendibility and Effectiveness as quality attributes.

To analyze a quality attribute it is mapped with a set of object oriented design properties. Design properties can be measured or observed using functionality, relationship and structure of design components [11].

2.1.7 Maintainability Index (MI)

Maintainability index is combination of few metrics. It is combination of LOC, Volume proposed by Halstead and Cyclometic complexity metric by McCabe. It is calculated as:

$$MI = 171 - 5.2\ln(V) - 0.23V(g) - 16.2\ln(LOC) \qquad (1)$$

Here V is volume and V(g) is complexity [12].

Maintainability index can be used as quality indicator. Higher values show that it is easy to maintain code. Three ranges are defined high, medium and low for MI. 0–64 is range for low MI. 65–84 indication of medium maintainability and 85–118 as high maintainability [13].

2.1.8 Coupling and Cohesion Metrics

MPC

Message Passing Coupling metric depicts complexity of messages transferred among different classes. It can be calculated by calculating number of send statements. This metric indicates dependency of class methods on other classes.

DAC

Data Abstraction Coupling is a measure of coupling created through abstract data types. If a class A has a property of type B means A contains a property x of type class B, this is DAC coupling as class A can access all data and methods of class B via property x. DAC is calculated by counting number of ADTs used in a class [14].

TCC

Tight Class Cohesion measures the degree of cohesion of class based on direct connection among pairs of methods and maximum possible number of pairs of methods.

LCC

Loose Class Cohesion measures cohesion based on direct plus indirect connections among method pairs and maximum possible number of pairs of methods [15].

2.2 Software Quality Frameworks and Tools

Measuring source code quality is a challenge and many researchers have worked on proposing different quality measurement frameworks. Also, there are many tools available that calculate different code metrics to provide quality information about source code. Some of the frameworks and tools are discussed in this section that used metrics to calculate source code quality.

2.2.1 Source Code Quality Framework for C Language (SCQFC)

This framework was proposed to analyze code quality of C language-based programs. This framework provides analysis of portability, maintainability, reliability and reuse ability of code in quantitative form. This framework used QAC and LogiScope tools for calculating code metrics. Although this framework provides quality analysis in terms of quality factors specified by ISO9126 but it is not making use of most popular metrics proposed in literature for object-oriented code [16].

2.2.2 Intelligence Code Evaluator (ICE)

Intelligence Code Evaluator is tool for Java source code for analyzing code quality based on metric values. Sequencer, syntax analyzer, metric analyzer and evaluator are basic components used to analyze Java code. This tool used only few metrics from literature to measure code quality [17].

2.2.3 Designite

Designite is code evaluation tool that analyses quality through code smells which appear at design level. This tool is implemented in C#. Code is parsed via another tool named NRefactory. NRefactory creates AST from Parsed code. Proposed tool use AST to create a meta-model of hierarchical type. This meta-model contains project objects. Project object has namespace objects that are part of project. And namespace object contains class/type objects [18].

2.2.4 QualityGate SourceAudit

SourceAudit tool measures the maintainability of software code using standards defined in ColumbusQM model of ISO/IEC 25010. This tool measures the maintainability of source code using metrics and then aggregating metrics to high level elements. If code has higher maintainability its development cost is low and vice versa. This tool analyses code using benchmarks [19].

2.2.5 PHP_depend

PHP_depend is PHP based static code analyzing tool. It analyses coupling, complexity, inheritance and size of software code. This tool shows results in form of pyramid. Inheritance metrics are shown on top, right half contains coupling metrics and left half shows size and complexity metrics. This tool is quite useful but it does not depict quality at higher level [20].

2.2.6 CodeMR

CodeMR is tool for static analysis of code for Java, Scala and C++. It uses code metrics and quality attributes for evaluation of code. It measures coupling, complexity and size of software code [21].

2.2.7 PHPMetrics

This tool is developed in PHP to analyze source code written in PHP. It uses various procedural and object-oriented metrics to calculate complexity and instability of code. The results are represented visually to make analysis easy [22].

3 Methodology

This study proposes a framework OOCQM for depicting source code quality in numerical form with details in terms of quality factors. This framework is particularly proposed for object-oriented code which is most popular coding paradigm as well as most studied code type in literature in terms of quality measurement.

Software quality models like Bohme model, McCall model and ISO9126 divided quality in factors. Like McCall model divided quality in reusability, portability, interoperability, reliability, correctness, efficiency, testability, integrity, usability, maintainability and flexibility.

Source code quality is a bit different from general software quality. Quality factors declared suitable for measuring source code quality are:
Reliability, Maintainability, Testability, Reusability, Portability, Understandability, Simplicity, Auditability [23].

A study conducted in 2016 for analyzing the relationship between code quality factors and code quality metrics. This study found that most of the quality factors can be mapped with few metrics proposed for OO code. Mapping results of this study are used as basis for this framework. Mappings of source code metrics against quality factors are shown in Table 1.

Some of the metrics are common for some quality factors. Arvanitou et al. [24] during the process of detailed analysis of mapped metrics, definition or calculation

Table 1 Quality factors and source code metrics mappings

Code quality factors	Source code metrics
Maintainability	DIT, LOC, WMC, CC-VG, TCC, NOCC, RFC, MPC, DAC, NOM
Reusability	LCOM, LOC, CBO, RFC, MPC, WMC, NOCC
Change proneness	DIT, NOCC, CBO, RFC, LCOM, DAC, NOA
Stability	WMC, LOC
Testability	RFC, CBO, LCOM, LOC

formula of some metrics cannot be traced so those metrics are not used in framework like ECC, ECS etc. Only five factors maintainability, re usability, change proneness, testability and stability are chosen. Modifiability and understandability also have mappings for some code metrics. But these two factors are not selected as these are sub factors of maintainability and all metrics in these two factors are also mapped with maintainability. Testability is also a sub factor of maintainability but it is considered as separate factor in proposed framework as it has one metric different from maintainability and also due to the importance of testability in code quality.

3.1　OOCQM Framework

Proposed framework is shown in Fig. 1. It consists of four components.

1. Read and Parse Code
2. Metrics Calculation and Normalization
3. Quality Factors Calculation
4. Quality Calculation.

Fig. 1 OOCQM framework

3.1.1 Read and Parse Code

This component reads object-oriented code of selected project. Size can vary from few classes to hundreds of classes/files. A language specific parser (Nikic for PHP) is used to parse code to generate Abstract Syntax Tree (AST). AST allows calculation of metrics in a convenient way.

3.1.2 Metrics Calculation and Normalization

Definition of every selected metric is traced from literature and then calculated according to definition. A database is used to save intermediate values that are used in some metric calculations. Database also saves metric values based on class, based on quality factor and based on project.

Different metric values have different ranges like DIT has value n where n can be any positive integer including zero. While value of metric TCC lies between 0 and 1. Metric values are evaluated in most studies through thresholds. But thresholds approach has following shortcomings:

- Thresholds are not available for all metrics.
- Threshold values for one metric vary in different studies so difficult to choose one threshold.
- Threshold values are dependent on other factors like code size, code type etc.

This study suggests to normalize each metric value using min max normalization technique. First of all, metric values of all classes are calculated then these values are normalized using minimum and maximum value of that metric. Let's say there are N classes in a project and M is set of all values of a metric m. Normalized Value V' of a class C calculated as:

$$V'(Cm) = V - \max(M) \big/ \max(M) - \min(M) \qquad (2)$$

Here V is the actual value of metric m for class C. This method makes all values of all metrics fall in range of 0–1.

After calculating and normalizing metrics values for all classes. Project base metric value is calculated by calculating average metric value of all classes. For a project P metric value calculation can be described as show in equation below. Here n is count of classes in project. Cm is metric value for class C.

$$V(Pm) = \frac{\sum_{i=1}^{n}(C_m)}{n} \qquad (3)$$

3.1.3 Quality Factors Calculation

Five quality factors are chosen for OOCQM. Only those quality factors are selected for which code metrics exist in literature. Each factor assigned 20 points so the sum of all factors is 100. All selected factors maintainability, change proneness, reusability, testability and stability are given equal points as all of these are equally important for a good quality software. The value of each quality factor is calculated based on all metric values that are mapped against that factor. It is tested that all metrics that are mapped against a quality factor has negative correlation with selected factor. For example, let's have look at metric WMC. It is mapped against maintainability, reusability and stability. Increase in value of WMC decreases these three factors. To create a relationship between metrics and quality factors that can depict this negative correlation following equation is devised.

$$F = \sum_{i=1}^{n} (20/n) * (1 - V) \tag{4}$$

Here F is quality factor value, n is number of mapped metrics against selected factor. V is average value of metric.

3.1.4 Quality Calculation

Calculated values of all five quality factors are aggregated up to make a numerical value between 1 and 100. It is described in Eq. 5 below. In this equation Q is quality of project. N is count of quality factors in framework. f (v) is value of selected factor

$$Q = \sum_{i=1}^{n} (f(v)) \tag{5}$$

3.2 OOCQM Implementation

This framework is implemented as tool to validate the proposed framework. Tool is implemented in PHP for evaluation of OO PHP code. Nikic parser is used in this tool for code parsing. AST are generated from parsed code. Software code quality metrics are calculated from generated ASTs. Some metrics are calculated directly from parsed code like LOC or NOM. Some metrics calculation used database for intermediate variable calculations. This tool is developed in Laravel.

Calculation formulas for selected metrics are picked from previous studies where these metrics were proposed. Metrics are calculated at class level. To cater the variation in code size that is lines of code of a class metric values are normalized using min-max normalization between 0 and 1 inclusive. Normalized metric values of all classes of a project are then added and result is divided by total number of classes to get a metric value at project level.

4 OOCQM Evaluation

Validation of proposed framework is very important because without validation it is not possible to prove its usefulness for research community and software industry and academics. Laravel is considered one of the best PHP frameworks for developing web applications. Five versions of Laravel are chosen to apply OOCQM. Results of OOCQM are shown in Table 2.

Results of OOCQM are compared with MI as shown in Table 2. We have evaluated that quality values generated by OOCQM are in accordance with MI values for different versions of Laravel. MI values of higher than 118 are considered very good. This suggests that all selected version of Laravel have good quality.

We compared OOCQM with some other quality analysis tools s shown in Table 3. Most of these tools are not calculating code quality in terms of quality factors. SCQFC measures quality in terms of reliability, maintainability and reusability but it does not calculate quality as a numerical value. PhpMetrics calculates only maintainability of source code.

5 Conclusion and Future Work

OOCQM is a generic code quality analyzer for object-oriented code. It consists on popular metrics some of which are never used collectively before in any tool for quality analysis of source code. This framework measures different aspects of

Table 2 OOCQM results

Project	Maintainability	Reusability	Change proneness	Testability	Stability	Quality	MI
Laravel 5.1	12.75	14.03	12.63	12.99	10.27	62.68	194.95
Laravel 5.2	11.47	13.69	11.83	11.7	11.97	60.69	193.53
Laravel 5.3	12.16	13.18	11.92	11.92	10.2	59.63	191.3
Laravel 5.4	12.41	13.48	12.15	12.32	10.06	60.45	192.6
Laravel 5.5	12.45	13.76	11.82	12.54	10.44	61.02	193.5

Table 3 OOCQM comparison with other tools

Tool name	Maintainability	Testability	Reusability	Change proneness	Stability	Quality
SCQFC [16]	Y	N	Y	N	N	N
Designite [18]	N	N	N	N	N	Y
PHP_depend [20]	N	N	N	N	N	N
SourceAudit [19]	Y	N	N	N	N	N
ICE [17]	N	N	N	N	N	Y
OOCQM	Y	Y	Y	Y	Y	Y
CodeMR [21]	N	N	N	N	N	Y
PHPMetrics [22]	Y	N	N	N	N	Y

object-oriented code like Size, complexity, cohesion, coupling and abstractness of code and converts these measurements into numerical value of quality. It calculates quality factors which are not calculated before using metrics. We also used some metrics like RFC, DAC in calculation of quality factors which are never used in any framework previously. In future it can be extended for more programming languages. Quality value threshold is an aspect that can be further explored that which quality number is good and which is not good.

References

1. Cavano, J.P., McCall, J.A.: A framework for the measurement of software quality. In: ACM SIGSOFT Software Engineering Notes, vol. 3, no. 5, pp. 133–139 (1978)
2. Kitchenham, B., Pfleeger, S.L.: Software quality: the elusive target. IEEE Softw. **13**(1), 12–21 (1996)
3. Deissenboeck, F., Juergens, E., Lochmann, K., Wagner, S.: Software quality models: purposes, usage scenarios and requirements. In: Proceedings of International Conference on Software Engineering, pp. 9–14 (2009)
4. Nuñez-Varela, A.S., Pérez-Gonzalez, H.G., Martínez-Perez, F.E., Souberville-Montalvo, C.: Source code metrics: a systematic mapping study. J. Syst. Softw. **128**, 164–197 (2017)
5. Boehm, B.W., Brown, J.R., Lipow, M.: Quantitative evaluation of software quality. In: Proceedings of 2nd International Conference on Software Engineering, pp. 592–605 (1976)
6. Rosenberg, J.: Some misconceptions about lines of code. In: Proceedings of Fourth International Software Metrics Symposium, pp. 137–142 (1997)
7. Of, A., Designs, T.: Halstead's Metrics: Analysis of Their Designs, vol. 1, pp. 145–159 (1977)
8. McCabe, T.J.: A complexity measure. IEEE Trans. Softw. Eng. **SE-2**(4), 308–320 (1976)
9. Chidamber, S.R., Kemerer, C.F., Chidamber, S.R., Kemerer, C.F.: Towards a metrics suite for object oriented design. In: ACM SIGPLAN Notices, vol. 26, no. 11, pp. 197–211 (Nov 1991)

10. Harrison, R., Counsell, S.J., Nithi, R.V.: An evaluation of the MOOD set of object-oriented software metrics. IEEE Trans. Softw. Eng. **24**(6), 491–496 (1998)
11. Bansiya, J., Davis, C.G.: A hierarchical model for object-oriented design quality assessment. IEEE Trans. Softw. Eng. **28**(1), 4–17 (2002)
12. Welker, K.D.: The software maintain ability index revisited. CrossTalk **14**, 18–21 (2001)
13. Anggrainingsih, R., Johannanda, B.O.P., Kuswara, A.P., Wahyuningsih, D., Rejekiningsih, T.: Comparison of maintainability and flexibility on open source LMS. In: Proceedings of 2016 International Seminar on Application for Technology of Information and Communication. ISEMANTIC 2016, pp. 273–277 (2017)
14. Li, W., Henry, S.: Object-oriented. J. Syst. Softw. **23**(2), 111–122 (1993)
15. Ott, Linda M., Bieman, James M.: Program slices as an abstraction for cohesion measurement. Inf. Softw. Technol. **40**(11-12), 691–699 (1998)
16. Washizaki, H., Namiki, R., Fukuoka, T., Harada, Y., Watanabe, H.: A framework for measuring and evaluating program source code quality. In: International Conference on Product Focused Software Process Improvement, pp. 284–299 (2007)
17. Sangeetha, M., Arumugam, C., Senthil Kumar, K.M., Alagirisamy, P.S.: Enhancing internal quality of the software using intelligence code evaluator. In: Communications in Computer and Information Science (CCIS), vol. 330, pp. 502–510 (2012)
18. Sharma, T.: Designite—A Software Design Quality Assessment Tool, pp. 1–4 (2016)
19. Bakota, T., Ladányi, G., Ferenc, R.: QualityGate Source Audit: A Tool for Assessing the Technical Quality of Software, pp. 440–445 (2014)
20. Randriatoamanana, R.: Object Oriented Metrics to measure the quality of software upon PHP source code with PHP_depend study case request online system application, pp. 2–6 (2017)
21. CodeMR Follow: CodeMR Static Code Analysis Tool. LinkedIn SlideShare, 14 June 2018. www.slideshare.net/codemr/codemr-software-quality
22. Lépine, J.-F.: PhpMetrics: Static Analyzys for PHP. Of Php. www.phpmetrics.org/documentation/index.html
23. Iqbal, T., Iqbal, M., Asad, M., Khan, A.: A Source Code Quality Analysis Approach
24. Arvanitou, E.M., Ampatzoglou, A., Chatzigeorgiou, A., Galster, M., Avgeriou, P., Arvanitou, E.M.: PT US CR. J. Syst. Softw. (2017)

A Fault-Tolerant and Flexible Privacy-Preserving Multisubset Data Aggregation in Smart Grid

Hung-Yu Chien and Chunhua Su

Abstract Smart Grid (SM) facilitates the intelligent generation, management, and distribution of electricity. It will be a very important service in our daily lives, and the security and privacy protection of the information and the structure is critical. Privacy-Preserving Data Aggregation (PPDA) in smart grids aims at collecting the aggregated power generation or consumption while protecting the privacy of each individual Smart Meter (SM). Li et al.'s Privacy-Preserving Multisubset data Aggregation (PPMA) (Li et al. in IEEE Trans Ind Inf 14(2):462–471, 2018 [1]) is at the cutting edge of PPDA schemes. Li et al.'s PPMA scheme, in addition to the total aggregated electricity, further provides the number of users whose electricity consumptions fall within an interested range and the aggregated quantity of the specified range. However, the requirement of strict time synchronization and no single SM failure makes the scheme un-attractive to practical application. We propose a new PPMA scheme that facilitates flexible SM deployment, independent SM status reporting without strict synchronization, and fault tolerance to any SM failure as long as at least two well-function SMs.

Keywords Smart grid · Privacy · Encryption · Aggregation · Smart meter

1 Introduction

Smart grids empower the traditional power grids with the capacities of information processing and data communications to facilitate reliable and efficient electricity generation, consumption, transmission, distribution, and control [2, 3]. Smart Meters (SM) deployed at the user side record the information and periodically report the

H.-Y. Chien (✉)
Department of Information Management, National Chi-Nan University,
470 University Road, Puli, Nantou, Taiwan, R.O.C.
e-mail: hychien@ncnu.edu.tw

C. Su
Division of Computer Science, The University of Aizu, Aizuwakamatsu, Japan
e-mail: chsu@u-aizu.ac.jp

© Springer Nature Switzerland AG 2020
R. Lee (ed.), *Computational Science/Intelligence and Applied Informatics*,
Studies in Computational Intelligence 848,
https://doi.org/10.1007/978-3-030-25225-0_12

data. Based on the collected data, a Control Center (CC) in smart grids can improve dynamic demand response and energy use.

The information collected by SMs should be securely transmitted and the privacy of individual information should be properly protected, since the data reported by the SMs would reveal the real-time usage and the related behaviors of the users. To preserve the privacy of transmissions and of the users, one straightforward approach is to encrypt the individual data and separately transmitting the encryptions; but this approach will incur heavy computations or communications on the core network when there are lots of SMs reporting, and it cannot protect the privacy of individual user. One another promising approach is the Privacy-Preserving Data Aggregation (PPDA) [1, 4–14] that aggregates the encryptions such that the communication overhead is reduced, CC can derive the aggregated power consumption in a geographical region while protecting users' fine-gained data from disclosure. Paillier homomorphic cryptosystem [15], lattice-based mechanism [9] or ElGamal-based mechanism [10] are popular building blocks for designing PPDAs.

However, existent PPDA schemes [1, 4–10, 13, 14] can only report the aggregated power consumption of a set of users, but not any information of the number of users whose consumptions fall within a specified range. In order to have better power generation prediction and planning, it is desirable that a scheme can not only report the aggregated consumption but also the number of users whose electricity consumptions are within a specific range. To tackle this challenge, Lu et al. scheme [11] can aggregate two separated two sets, and Lu et al. [12] further designed the EPPA scheme to allow flexible sets aggregations.

Recently, Li et al. propose a promising Privacy-Preserving Multi-subset Aggregation (PPMA) scheme [1], in which, the users set in a residential area are divided into multisubset according to their electricity consumptions in each period, and CC can obtain the sum of electricity consumption and the number of users for each subset. The PPMA scheme can protect each user's privacy from the collusion of the CC and the GW. However, we notice that, even though there exist many PPDA schemes and PPMA schemes, there are still several critical weaknesses that keep them from practical use. In this paper, we will critically discuss the desirable features and functions for further PPDA/PPMA research, and propose a new PPMA scheme. The proposed PPMA scheme can achieve the desirable features and improve the efficiency. The rest of this paper is organized as follows. Section 2 discusses the desirable features of practical privacy-preserving data aggregation for smart grids. Section 3 proposes our new PPMA scheme. Section 4 analyzes the security properties, and evaluates the performance. Section 4 analyzes the security properties and the efficiency. Section 5 states our conclusions.

2 Discussions of Requirements

Even though Li et al.'s PPMA and the related works have improved the functions and efficiency of PPDA/PPMA schemes to some extent, there are several critical criteria and functions being neglected by existent schemes. These weaknesses will deter the practical applications of these schemes and limit the perspectives of further research. Therefore, we examine these critical criteria for future research. We respectively discuss these criteria and weaknesses inherited in these schemes as follows.

Criterion 1: Flexible SM deployment

Smart grids should efficiently adapt to the dynamic change of SM deployment, as users dynamically join or leave the communities. Regarding this feature, any PPDA and PPMA schemes should support dynamic user/SM deployment. However, in Li et al.'s PPMA scheme and some others, each $SM_i \in \{SM_1, \ldots, SM_n\}$. is assigned a secret value x_i and CC is assigned a secret value x_0 such that they should satisfy the condition $x_0 + x_1 + \cdots + x_n = 0 \bmod N$ for the aggregated encryption to be properly decrypted by the CC. Unfortunately, if any membership changes, then the condition does not hold and the secret values should be re-assigned and re-distributed.

Criterion 2: Independent SM status reporting without strict synchronization

As the SMs are expected to report their recorded data in real-time and frequently, it is required that these SMs should be able to report their data independently without strict synchronization; otherwise, any possible problem in the synchronization would worsen the efficiency or even cause the system malfunction. For those encryptions to be aggregated in Li et al.'s PPMA, these SMs should synchronize their timestamp t and use the same t in calculating $c_i = g^{a_j \Delta m_i} g^{b_j} H(t)^{x_i N} \bmod N^2$. This requirement of synchronization would seriously incur extra overhead and even fails the system, due to any network problems or disturbances.

Criterion 3: Fault tolerance to any SM failure

Since a Gateway is expected to aggregate several SMs under its domain, a PPDA scheme or a PPMA scheme should be tolerant to any possible SM failure; a single or several SM failures should not deter the rest of the smart meters from reporting their data. However, in Li et al.'s scheme, the CC would require $x_0 + \cdot \sum_{i=1}^{n} x_i \bmod N$ *being zero* when all SMs function properly to and correctly contribute their shares; if any one of these SMs cannot function properly, then $x_0 + \cdot \sum_{i=1}^{n} x_i \bmod N = 0$ does not hold and the aggregation fails. That is, Li et al.'s PPMA scheme cannot tolerate any SM failure.

Criterion 4: The data expansion of encryption should be modestly efficient

Considering that smart meters are not powerful devices and the communication channels are precious resources, the design of PPDA and PPMA should seriously take transmission efficiency and computation efficiency into account. Regarding the transmission efficiency of single encryption, we concern how the length of the message expands after encryption. We define the term Transmission Efficiency of Single plaintext (TES) by calculating the ratio between the bit length of the plaintext m_i (which is bounded by the maximum E) and the bit-length of its encryption; we define TES $= \frac{|a\ plaintext|}{|a\ ciphertext|}$.

Criterion 5: The transmission overhead reduction of the aggregated encryption should be as large as possible

In addition to examine how an aggregated encryption improves the efficiency of aggregation, PPDA/PPMA scheme should try to achieve as greater efficiency as possible after aggregating many encryptions. Regarding this goal, we define the term Transmission Efficiency of Aggregating encryptions (TEA) by calculating the ratio between the total bit length of n plaintexts and the bit-length of their aggregated encryption; we define TEA $= \frac{n * |plaintext|}{|aggregated\, ciphertext|}$.

Criterion 6: The computation on the smart meter should be modestly efficient and affordable

Even though smart meters are not resource-constrained devices like Radio Frequency Identification tags or sensors, those expensive computations like pairing operations should be avoided to reduce the computation loading.

3   New PPMA Scheme

Our PPMA scheme is based on Paillier cryptosystem. Paillier cryptosystem is reviewed in Sect. 3.1, and the proposed scheme is introduced in Sect. 3.2.

3.1   *Paillier Cryptosystem [15]*

Paillier cryptosystem consists of the three phases: key generation, encryption, and decryption.

(1) *Key Generation*: Select two large and independent prime numbers p and q randomly, and set $N = p * q$. Define $L(x) = (x-1)/N$ and $g = (1 + N)$. Compute $\lambda = lcm(p - 1, q - 1)$ and $\mu = \left(L\left(g^\lambda\, mod\, N^2\right)\right)^{-1} mod\, N$, where lcm stands for the least common multiple. The public key is (N, g), and the private key is (λ, μ).

(2) *Encryption*: Given a message $m \in Z_N^*$, it selects a random number $r \in Z_{N^2}^*$, and then the ciphertext can be computed as follows.

$$c = E(m) = g^m r^N\, mod\, N^2 \tag{1}$$

(3) *Decryption*: Let c be the ciphertext to be decrypted, where $c \in Z_{N^2}^*$, the plaintext is computed as $m = L\left(c^\lambda\, mod\, N^2\right) \cdot \mu\, mod\, N$. The correctness of the formula can be verified as follows, and the interested readers are referred to [15] for the detailed proof of the correctness.

$$c^\lambda = g^{\lambda m} r^{\lambda N} = g^{\lambda m} r^{\emptyset(N^2)} = g^{\lambda m} = (1 + N)^{\lambda m} \bmod N^2 \tag{2}$$

We can expend the power $(1 + N)^m$ with Binomial theorem as follows.

$$(1 + N)^{m\lambda} = \sum_{i=0\sim m} \binom{m\lambda}{i} N^i$$
$$= 1 + m\lambda N \bmod N^2, \text{ as all items with } i \geq 2 \text{ turn to zero.} \tag{3}$$

3.2 The Proposed PPMA

The proposed PPMA has four phases, (1) system setup, (2) encrypt electricity consumption, (3) aggregate the encryptions, and (4) decrypt the aggregation.

(1) *System setup*: We make a little twist to the Paillier cryptosystem's parameters as follows.

TTP sets up the public key $N = pq$, $g = (1 + N)$, $H()$ a hash function, $\lambda = lcm(p - 1, q - 1) = \lambda' w_1 w_2$, where w_1, w_2 are two small factors of λ. Assume there are maximum n SMs' encryptions to be aggregated in a GW's domain, and the electricity consumption is divided into k ranges—$[R_1, R_2), [R_2, R_3), \ldots, [R_k, E]$. Each smart meter SM_i is assigned a secret key x_i. Define $b_j = (n + 1)^{j-1}$, $j = 1 \sim k$. The system-wise public key includes $(N, g, w_1, w_2, n, k, \{b_j s\})$, and the private key includes (λ, λ', μ).

(2) *Encrypt electricity consumption*:

Let U_i be a user $U_i \in U = \{U_1, \ldots, U_n\} = \overline{U_1} \cup \overline{U_2} \cup \cdots \cup \overline{U_k}$. Assume U_i has his consumption $m_i \in [R_j, R_{j+1})$, then he computes $c1_i, c2_i$ as follows, where t_i is the timestamp of SM_i. Then, he sends $c1_i, c2_i$ to the GW.

$$c1_i = g^{mi} \cdot H(t_i)^{Nw_1w_2x_i} \bmod N^2 \tag{4}$$

$$c2_i = g^{bj} \cdot H(H(t_i))^{Nw_1w_2x_i} \bmod N^2 \tag{5}$$

(3) *Aggregate the encryptions*:

Upon receiving the encryptions from its users, GW performs the aggregation as follows:

$$C1 = \prod_{j=1\sim k, U_i \in \overline{U_J}} c1_i = \prod_{j=1\sim k, U_i \in \overline{U_J}} g^{mi} \cdot H(t_i)^{Nw_1w_2x_i}$$
$$= g^{\sum mi} \cdot \prod_{j=1\sim k, U_i \in \overline{U_J}} H(t_i)^{Nw_1w_2x_i} \bmod N^2 \tag{6}$$

$$C2 = \prod_{j=1\sim k, U_i \in \overline{U_J}} c2_i = \prod_{j=1\sim k, U_i \in \overline{U_J}} g^{bi}.$$

$$= g^{\sum bi} \cdot \prod_{j=1\sim k, U_i \in \overline{U_J}} H(H(t_i))^{N w_1 w_2 x_i} \bmod N^2 H(H(t_i))^{N w_1 w_2 x_i} \qquad (7)$$

4) *Decrypt the aggregation*:

CC first computes V_1 and V_2, and then derives the total consumption and B as follows.

$$V_1 = C1^{\lambda'} = g^{\lambda' \sum m_i} \cdot \prod_{j=1\sim k, U_i \in \overline{U_j}} H(t_i)^{\lambda' N w_1 w_2 x_i} = g^{\lambda' \sum_{j=1\sim k, U_i \in \overline{U_J}} m_i} \qquad (8)$$

$$V_2 = C2^{\lambda'} = g^{\lambda' \sum b_i} \cdot \prod_{j=1\sim k, U_i \in \overline{U_J}} H(H(t_i))^{\lambda' N w_1 w_2 x_i} = g^{\lambda' \sum_{j=1\sim k, |\overline{U_J}| * b_j}} \qquad (9)$$

$$\sum_{j=1\sim k, U_i \in \overline{U_J}} m_i = (V_1 - 1)/(\lambda' N) \bmod N \qquad (10)$$

$$B = \sum_{j=1\sim k,} \overline{|U_J|} * b_j = (V_2 - 1)/(\lambda' N) \bmod N \qquad (11)$$

To derive the number of users in each range, we just execute the following Algorithm 1. The aggregated electricity consumption for each range $[R_i, R_{i+1})$ can be estimated as $\left|\overline{U_{iota}}\right| * \frac{(R_i + R_{i+1})}{2}$.

Algorithm 1: Recover(B)

For $i=k$ to 1 do

   $|\overline{U_\iota}| = (B - B \bmod b_i)/b_i$

   $B = B - (b_i \cdot |\overline{U_\iota}|)$

End for

   Return $\{|\overline{U_1}|, |\overline{U_2}|, ..., |\overline{U_k}|\}$

---

4 Security Analysis and Performance Evaluation

4.1 Security Analysis

The correctness

We first examine the correctness of the proposed scheme. The correctness of most steps of our scheme is obvious and easy to check. Now we check the correctness of (10) and (11). From (8), we have $V_1 = C1^{\lambda'} = g^{\lambda' \sum m_i} = (1+N)^{\lambda' \sum m_i} = \sum_{i=0 \sim \lambda' \sum m_i} \binom{\lambda' \sum m_i}{i} N^i = 1 + \lambda' N \sum m_i \bmod N^2$. So we have $\sum m_i = (V_1 - 1)/(\lambda' N) \bmod N$. This proves the correctness of (10). The same deduction can be applied on (9) to derive the result of (11).

The correctness of Algorithm 1 can be verified as follows. From (11), we have $B = \sum_{j=1 \sim k, U_i \in \overline{U_J}} b_i$. Because $b_i = (n+1)^{i-1}$ for $i = 1 \sim k$, B can be viewed as $B = \sum_{j=1 \sim k, U_i \in \overline{U_J}} b_j = \sum_{j=1 \sim k} \overline{|U_J|} b_j = \sum_{j=1 \sim k} \overline{|U_J|} (n+1)^{j-1}$. Since each $\overline{|U_J|} <= n$, it is easy to check that Algorithm 1 can easily derive each $\overline{|U_J|}$.

In the rest of this section, we analyze the security properties of our scheme.

Theorem 1 *The scheme protects privacy of individual electricity consumption and the aggregated electricity from outsiders.*

Proof An outsider might eavesdrop the transmissions (4) and (5) from the SM-GW channel and the transmissions (6) and (7) from the GW-CC channel. The individual $c1_i = g^{m_i} \cdot H(t_i)^{N w_1 w_2 x_i} \bmod N^2$ and the aggregated $C1 = g^{\sum m_i} \cdot \prod_{j=1 \sim k, U_i \in \overline{U_J}} H(t_i)^{N w_1 w_2 x_i} \bmod N^2$ contain the randomized $H(t_i)^{N w_1 w_2 x_i}$, which keeps the outsider from deriving the content. The individual $c1_i$ and the aggregated $C1$ are legal ciphertext of Paillier cryptosystem. As Paillier cryptosystem is semantic secure against the chosen plaintext attack, an outsider A is not able to recover the plaintext. The same results can be derived for $c2_i$ and $C2$. This proves the theorem. ∎

Theorem 2 *The collusion of several smart meters cannot violate the privacy of other uncorrupted smart meters.*

Proof From (4) to (7), we can see that each individual SM_i applies independent random number x_i in the encryptions. Therefore, any collusion of several meters would not endanger the privacy of other un-corrupted meters. ∎

Theorem 3 *The gateway GW cannot derive any individual electricity or the aggregated electricity consumption.*

Proof The gateway does not have more data or more capacities than an outsider. Following Theorem 1, it cannot derive either the individual electricity or the aggregated electricity. ∎

4.2 Performance Evaluation

We first evaluate the message expansion and the efficiency after aggregation. For that purpose, we examine the condition (bound) between the message lengths between the plaintext and the cipher-text. For our scheme to correctly recover the messages, the following conditions should be satisfied.

$$\lambda' \sum m_i < \emptyset(N^2) = \lambda' w_1 w_2 N, \quad \sum m_i < \emptyset(N^2) = w_1 w_2 N \qquad (12)$$

$$\lambda' \sum b_i < \emptyset(N^2) = \lambda' w_1 w_2 N, \quad \sum b_i < \emptyset(N^2) = w_1 w_2 N \qquad (13)$$

To further evaluate the message bit length bound, we assume all users consume the maximum electricity E. Then, we have the following.

$$\sum m_i < nE < w_1 w_2 N,$$

$$E < w_1 w_2 N/n, \log(E) < \log w_1 + \log w_2 + log N - \log(n) \qquad (14)$$

$$\sum b_i < nb_k = n(n+1)^{k-1} < w_1 w_2 N, \quad n(n+1)^{k-1} < w_1 w_2 N \qquad (15)$$

Equation (14) specifies the bound between the bit lengths of the plaintext and of the cipher-text, and it depends on the parameter settings and the value n. Equation (15) specifies the relation among the parameter setting (n, k, N). Based on Eq. (14), we now evaluate the TES of our scheme by calculating the ratio between the bit length of the plaintext m_i and the bit-length of its encryptions. Our TES is $\frac{|E|}{|C1|+|C2|}$. The TES of Li et al.'s PPMA is $|E|/|N^2|$.

Table 1 shows the bound of |E|, the TES, and the TEA for a practical setting of Li et al.'s PPMA. In the table, |N| = 1024 (N with bit-length being 1024), n = 1024 and 2048 (1024 and 2048 users), and k ranges from 4 to 102. For such a setting, the bit-length of Li et al.'s PPMA encryption is 2048 bits. The third row |E| shows the maximum bit length of |E| for that setting. When the electricity range is 100, the maximum plaintext bit-length is 50, the TES is 0.02, and the TEA is 25.

Table 2 lists the bound of |E|, TES and TEA for various (k, n) settings of our scheme. For $n = 1024$ and |N| = 1024, we have $\log(E) < \log w_1 + \log w_2 + 1024 - 10$. We approximate the above equation by letting w_1 and w_2 as small as 2. Then, we have $\log(E) \cong 1016$. That is, for a modulus N^2 with 2048 bits, it can support the maximum electricity consumption E with bit length up to 1016. So we have TES $= \frac{|E|}{|C1|+|C2|} = \frac{1016}{4096} = 0.248$ and the TEA is 254.6. For $n = 2048$ and $k = 80$, we have |E| = 1015, the TES is 0.248, and the TEA = 508.7. From Tables 1 and 2, we can see that, when k is greater than 100 and $n = 1024$, our scheme has better

Table 1 Bit-length comparison between the electricity plaintext and its cipher-text of Li et al.'s scheme, where |N²| = 2048, n = 1024 or 2048, and k ranges from 4 to 100

n	1024	1024	1024	1024	2048	2048	2048	2048
k	4	40	100	102	4	40	80	90
|E|	1970	1050	50	10	1962	1390	290	70
TES	0.96	0.61	0.02	0.004	0.95	0.57	0.14	0.03
TEA	985	625	25	5	1962	1170	290	70

Table 2 Bit-length comparison of our scheme, where $|N^2| = 2048$, $n = 1024$ and k ranges from 4 to 100

n	1024	1024	1024	*1024*	2048	2048	2048	2048
k	4	40	100	102	4	40	80	90
$|E|$	1016	1016	1016	1016	1015	1015	1015	1015
TES	0.24	0.24	0.24	0.24	0.24	0.24	0.24	0.248
TEA	254	254	254	254	508	508	508	508

communication performance; for $n = 2048$ and k is greater than 80, our scheme has better communication performance than Lie t al.'s scheme. In a short summary, our scheme has better performance in terms of transmission efficiency when k becomes larger.

Now we examine the computational performance. Here, we concerns those expensive computations like modular exponentiation (denoted as T_{ME}) and modular multiplication (denoted as T_{MM}) on SMs, GWs, and CC. Table 3 summarizes the comparison among the related works. *Please notice that, in Li et al.'s publication, they wrongly less count the numbers of T_{ME} of their scheme and the related works, because there are two different bases (g and H(t)) in the calculation $g^{a_j \Delta m_i} g^{b_j} H(t)^{x_i N} \mod N^2$ and they cannot be counted as one exponentiation.* Therefore, our scheme need almost twice the number of exponentiations than Li et al.'s scheme. Table 3 summarizes the comparison of the computation performance and the supported features. In a short summary, our scheme owns better performance in terms of fault tolerance, flexible SM deployment, the elimination of synchronization, and better communication performance when the number of electricity ranges is larger. Even though our SM requires two more modular exponentiations and one more modular multiplication, the extra computation is affordable and insignificant, since these two operations are not expensive for smart meters.

Table 3 Computational performance of the related works

	SM	GW	CC	FSMD	IRWOTS	FTF	SLN
Li	$2T_{ME} + 1T_{MM}$	$(n-1)T_{MM}$	$1T_{ME} + 1T_{MM}$	x	x	x	x
Our	$4T_{ME} + 2T_{MM}$	$2(n-1)T_{MM}$	$2T_{ME}$	V	V	V	V

FSMD Flexible SM Deployment; *IRWOTS* Independent SM status Reporting WithOut Time Synchronization; *FTF* Fault Tolerance to any SM Failure; *SLN* Support for Larger Number of electricity. *Levels* x: No; V: Yes

5 Discussions and Conclusions

In this paper, we have discussed the requirements of practical privacy-preserving data aggregation for smart grids. These criteria include (1) flexible SM deployment, (2) independent SM status reporting without strict synchronization, (3) fault tolerance to any SM failure, (4) the data expansion of the encryption should be modestly efficient, (5) the transmission efficiency and the computation loading on the smart meter should be modestly efficient. We have proposed a new PPMA scheme. The analysis shows that the proposed scheme owns much better performance in terms of functionalities and communication overhead, especially when the system would like to support larger number of electricity levels. In the future, we will explore the challenges of supporting larger number of users and the computation performance improvement.

Acknowledgements This project is partially supported by the National Science Council, Taiwan, R.O.C., under grant no. MOST 107-2218-E-260-001, and Chunhua Su is supported by JSPS Kiban(B) 18H03240 and JSPS Kiban(C) 18K11298.

References

1. Li, S., Xue, K., Yang, Q., Hong, P.: PPMA: privacy-preserving multisubset data aggregation in smart grid. IEEE Trans. Ind. Inf. **14**(2), 462–471 (2018)
2. Gungor, V.C., Lu, B., Hancke, G.P.: Opportunities and challenges of wireless sensor networks in smart grid. IEEE Trans. Ind. Electron. **57**(10), 3557–3564 (2010)
3. Fang, X., Misra, S., Xue, G., Yang, D.: Smart grid—the new and improved power grid: a survey. IEEE Commun. Surv. Tuts. **14**(4), 944–980 (2012)
4. Garcia, F.D., Jacobs, B.: Privacy-friendly energy-metering via homomorphic encryption. In: Proceedings of International Workshop Security Trust Management, pp. 226–238 (2010)
5. Chen, L., Lu, R., Cao, Z.: PDAFT: a privacy-preserving data aggregation scheme with fault tolerance for smart grid communications. Peer-to-Peer Netw. Appl. **8**(6), 1122–1132 (2015)
6. Chen, L., et al.: MuDA: Multifunctional data aggregation in privacy preserving smart grid communications. Peer-to-Peer Netw. Appl. **8**(5), 777–792 (2015)
7. Yang, Q., et al.: A privacy-preserving and real-time traceable power request scheme for smart grid. In: Proceedings of IEEE International Conference Communication, pp. 1–6 (2017)
8. Lu, R., et al.: EPPA: an efficient and privacy-preserving aggregation scheme for secure smart grid communications. IEEE Trans. Parallel Distrib. Syst. **23**(9), 1621–1631 (2012)
9. Abdallah, A., Shen, X.: A lightweight lattice-based homomorphic privacy-preserving data aggregation scheme for smart grid. IEEE Trans. Smart Grid. (To be published). https://doi.org/10.1109/tsg.2016.2553647
10. Dong, X., et al.: An Elgamal-based efficient and privacy-preserving data aggregation scheme for smart grid. In: Proceedings of IEEE Global Communications Conference, pp. 4720–4725 (2014)
11. Lu, R., Alharbi, K., Lin, X., Huang, C.: A novel privacy-preserving set aggregation scheme for smart grid communications. In: Proceedings of IEEE Global Communications Conference, pp. 1–6 (2015)
12. Lu, R., et al.: EPPA: An efficient and privacy-preserving aggregation scheme for secure smart grid communications. IEEE Trans. Parallel Distrib. Syst. **23**(9), 1621–1631 (2012)

13. Yang, Z., Yu, S., Lou, W., Liu, C.: P2: Privacy-preserving communication and precise reward architecture for V2G networks in smart grid. IEEE Trans. Smart Grid 2(4), 697–706 (2011)
14. Jo, H.J., Kim, I.S., Lee, D.H.: Efficient and privacy-preserving metering protocols for smart grid systems. IEEE Trans. Smart Grid 7(3), 1732–1742 (2016)
15. Paillier, P.: Public-key cryptosystems based on composite degree residuosity classes. In: Proceedings of International Conference on the Theory Applications of Cryptographic Techniques, pp. 223–238 (1999)

Secure and Efficient MQTT Group Communication Design

Hung-Yu Chien, Xi-An Kou, Mao-Lun Chiang and Chunhua Su

Abstract To facilitate the successful deployments of the Internet of Things (IoT) applications, the support of secure and efficient communication protocol and architecture is inevitable. Owing to its lightweight and easiness, the Message Queue Telemetry Transport (MQTT) has become one of the most popular communication protocols in the Internet-of-Things (IoT). However, the security supports in the MQTT are very weak: it assumes the security support from the underlying Secure Sockets Layer (SSL). The weakness incurs several key drawbacks. One is the support of SSL capacities is a pressure for those resources-constrained devices. One another and very important one is the lack of the support of secure group communication. Without efficient and secure group communication support, the MQTT-based IoT systems would suffer from deteriorated computational and communication performance, especially when there are tons of IoT devices accessing the systems. In this paper, we design a secure MQTT group communication framework in which each MQTT application would periodically updates the group key and the data communication can be efficiently and securely encrypted by the group keys. Both our prototype system and the analysis show that our design can improve the performance of security, computation, and communication.

H.-Y. Chien (✉)
Department of Information Management, National Chi Nan University,
Puli, Taiwan
e-mail: hychien@ncnu.edu.tw

X.-A. Kou · M.-L. Chiang
Department of Information and Communication Engineering, ChaoYang University of
Technology, Taichung City, Taiwan
e-mail: kandy841011@gmail.com

M.-L. Chiang
e-mail: mlchiang@cyut.edu.tw

C. Su
Division of Computer Science, The University of Aizu, Aizuwakamatsu, Japan
e-mail: chsu@u-aizu.ac.jp

© Springer Nature Switzerland AG 2020
R. Lee (ed.), *Computational Science/Intelligence and Applied Informatics*,
Studies in Computational Intelligence 848,
https://doi.org/10.1007/978-3-030-25225-0_13

Keywords Transport layer issues · Security and privacy · MQTT · Internet of Things · Authentication · Group communication

1 Introduction

Various Internet of Things (IoT) applications has been penetrating every sectors of our daily life. It is estimated that there will be billions of IoT devices deployed soon. To facilitate tremendous amount of frequent IoT connections and data transmissions, one key element is the support of efficient IoT communication protocols. Among several IoT communication protocols, the Message Queue Telemetry Transport (MQTT) [33] is the most popular one, owing to its lightweight and easiness to use. There are many MQTT-based IoT applications deployed globally [1].

However, MQTT itself does not provide the security protection like authentication, integrity, and confidentiality. It assumes the use of Secure Sockets Layer (SSL) in the underlying layer. However, SSL demands more computational resources, and authenticating clients using SSL requires the deployment certificates to every IoT devices, which is very effort-demanding. Additionally, MQTT itself does not enforce some desirable security properties and functions like secure group communication. Without the support of secure group communication, a publisher should encrypt its message first and sends to the broker; the broker then decrypts the ciphertext, individually use each subscriber's session key to encrypt the message again, and sends all the encryptions to all the subscribers; finally, each subscriber decrypts the data. This process not only puts a great burden on the broker but also deteriorates the overall communication performance.

Therefore, in this paper, we will propose a MQTT group communication framework which facilitates the group key distribution and group communication. We implement the framework and evaluate the performance. Both the experiments on the prototype system and the analysis show that our design improve the performance in terms of the security properties, the computation, and the communication. This paper has the following contributions. (1) A proposed MQTT group communication framework that can effectively improve the security and easiness to use. (2) The experiments on the prototype system shows its improvements on the communication performance. (3) The analysis, based on the collected data on the prototype system, shows its great improvement when there is a large volume of devices.

The rest of this paper is organizedas follows. Section 2 discuss the related publications and platforms. Section 3 proposes our MQTT group communication framework. Section 3.2 describes our prototype implementation. Section 4 evaluates the performance of our design. Section 5 states our conclusions and some future works.

2 Related Work

Among several popular IoT transmission protocols MQTT [2], Advance Message Queuing Protocol (AMQP) [3], Constrained Application Protocol (CoAP) [4], Extensible messaging and presence protocol (XMPP) [5], and Data Distribution Service (DDS) [5], MQTT is the one of the most popular one in consumer IoT applications, owing to its lightweight and easiness to use. It has been ratified as the ISO standard (ISO/IEC 20922: 2016) [6] and the OASIS standard [7].

A MQTT system consists of a set of clients and a broker who acts as an intermediary among the clients. The message exchange among clients is based on the concept of "topic". There are two kinds of clients. One is publisher who send messages to a broker who forward the messages to those subscribers. The other is subscriber who subscribes the messages of a topic from the broker. Figure 1 depicts the MQTT architecture, where "pollution" and "election" represent two topics. To keep the MQTT protocol lightweight and easy, the MQTT standard itself does not specify how to secure the transmissions and the accesses; Instead, it suggests the use of SSL/TLS and AES/DES for client authentication and for message encryption in the SSL layer. This simple principle makes it so lightweight, but also incurs a lots of security threats and risks [8].

Several MQTT platforms like [9–13] and many publications such as [14–23] have addressed the security weaknesses in some ways. However, none of existent solutions solve all the security challenges, and the support of group communications in MQTT has been neglected.

Andy et al. [15] and Firdous et al. [16] respectively demonstrated several attack scenarios on the MQTT platforms and their vulnerability to the Denial-Of-Service (DOS) attacks. Chien and Chen [14] concentrated on evaluating the security vulnerability of several Arduino products [24–28] acting as MQTT clients. Espinosa-Arandaet al. [21] designed a specialized hardware to help an IoT device handle the SSL connection. This extra hardware solution is costly for many IoT deployments. Lesjak et al. [23] designed a specialized hardware called the meditator to be integrated with an IoT device and to help the device handle the TLS server authentication with a MQTT broker.

Shin et al. [17], based on the Mosquitto 1.4.9 platform [10] and the AugPAKE protocol [18], designed the AugMQTT platform which provides device authentication and establishes session keys between a client and the MQTT broker. Bhawiyuga

Fig. 1 The MQTT architecture

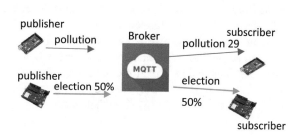

et al. [19] noticed the default authentication mechanism of using username and password in the MQTT API would have poor security and poor scalability; therefore, they propose their token-based authentication solution; however, the token just simply encodes the username and the password without using any encryptions, and there is no session key generation for the connections. Mektoubi et al. [20], based on the Public Key Infrastructure (KPI), system and the symmetric key encryption, design the client authentication and the topic-related message encryptions. One key feature of the scheme is that there is one specific certificate for each topic so that the messages for one topic can be encrypted using the public key of the certificate and can be decrypted using the corresponding private key. The private key is distributed to those subscribers. This feature facilitates the possible multicast of the topic messages. However, they also agree that the solution has several key weaknesses. One is the management of the lifecycle of the certificates/keys, and one another is the scalability challenge for a large number of client.

Rizzardi et al. [22] proposed a secure MQTT architecture of which the key management module is responsible for flexible key management for secure messages encryptions and key distributions, and a policy management module for users to specify the access policies of user-crate topics (applications). However, no specific algorithms are specified to fulfill the claimed functions. Chien et al. [29] systemically examine the security requirements of MQTT systems, and propose a security-enhanced MQTT platform where MQTT-API-compatible client authentication is emphasized. However, none of existent solutions provide secure and efficient group communications.

3 The Proposed MQTT Group Communication Architecture and the Prototype System

The proposed MQTT group communication architecture is proposed in Sect. 3.1, and the prototype system is introduced in Sect. 3.2.

3.1 The MQTT Group Communication Architecture

Figure 2 shows our MQTT group communication architecture. Our system is based on Chien et al.'s MQTT framework [29] and extends it with group communication. In Chien et al.'s framework, each device is required to be registered in the system, and each client should be authenticated before it can access the services. In each session, a client and its broker mutually authenticate each other, and they share one session key. In Fig. 2, we only show the group communication flows but not all the flows. In the group communication, there is one ManaGemenT server (MGT) which

Fig. 2 The MQTT group communication architecture

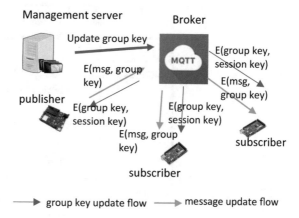

is responsible for periodical group key updating all the applications (they are also called "things" in several MQTT platforms).

When a user creates a thing (for example, say "humidity"), the MGT automatically creates a group-key-update thing (called updatekey/humidity), and all the publishers and subscribers of this thing automatically enrolled in this group-key-update thing. The flows marked in red color denote those flows for updating the group keys. First, the MGT periodically updates the group keys, and securely send the new group key to the broker. The broker then encrypts the group key, using each client's session key. The flows marked in blue color denote the normal MQTT messages. A publisher encrypts its messages using the group key, and sends the encrypted messages to the broker. The broker does not decrypt the encrypted and directly forwards the encryptions to all subscribers. The subscribers decrypt the encryptions, using the group keys.

In Fig. 3, it shows the normal MQTT flows and the group-key-update MQTT flows. The two kinds of flows follow after the Challenge-Response (C-R) authentications. The group-key-update flows are special MQTT flows in which only the MGT is the publisher.

3.2 The Prototype System

Based on the open source Mosca platform [13], web socket, JSON [30], Arduino [24], XMPP [32] and node.js [31], we have implemented the proposed MQTT group communication framework (depicted in Fig. 2). Here, we introduce some major functions of our extension of the Mosca platform. We respectively introduce these functions, based on the categories.

Figure 4a shows a client uses the group key to decrypt the received message. Figure 4b shows a broker received a group-key-update message from the MGT in the marked red rectangles, and the red rectangles shows the broker publishes the group-

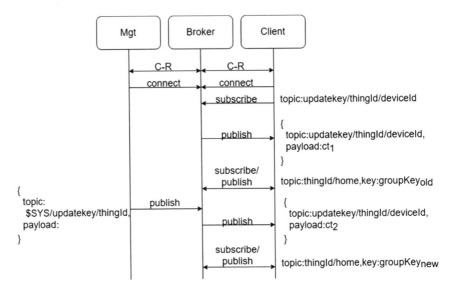

Fig. 3 The update-key MQTT flows and the normal MQTT flows

key-update message to a client. In Fig. 4b, the first blue rectangle "$SYS/updateKey" shows that it is a group-key-update message, the second blue rectangle "5f … 06" shows the specified thing identification. The first red rectangle identifies it is a group-key-update message for the thing with the identity "5f … 06", and the second red rectangle specifies the device identity for this message. The encoded message embraced in the yellow braces is the encrypted group key encrypted using the session key. Figure 4c shows two successive group-key-update messages. Figure 4d shows the logged messages of a subscriber. The red rectangles show the received group-key-update messages. The green rectangles show the group key used to decrypt the messages. The rest white texts show the decrypted messages.

4 Performance Evaluation

To evaluate the performance, we implement our system with the specified hardware and software in Table 1. We describe the experiment environments, and then discuss the performances. The experiment is conducted in a wired LAN to avoid possible communication disturbance of wireless links. The clients is run on the node, js platform.

We run two experiments. The two experiments all involve one publisher, three subscribers, and one broker. The publisher publishes one message to the broker, and then the broker forwards it to the subscribers.

In the first experiment, the publisher encrypts the message, using the session key; the broker decrypts it, using the session key; it then uses three different session

Fig. 4 Some messages from
the MQTT group
communications

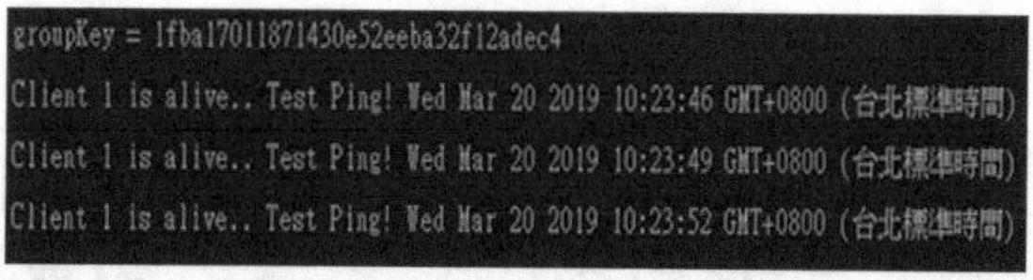

(a) Subscriber decrypts the message using the group

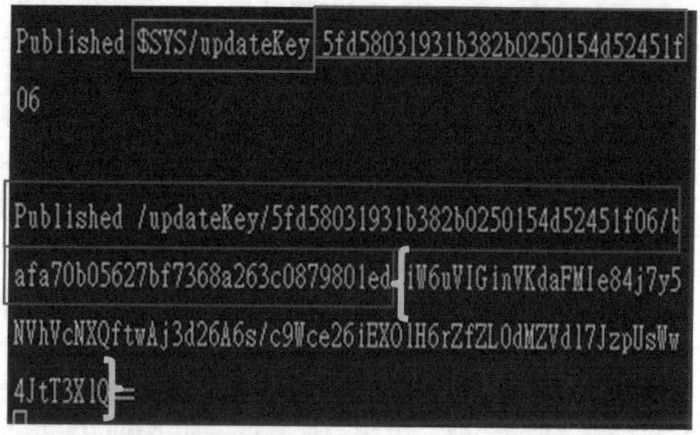

(b) A broker received a group-key-update message
and publishes it to a client

(c) One device gets two group-key-update
messages

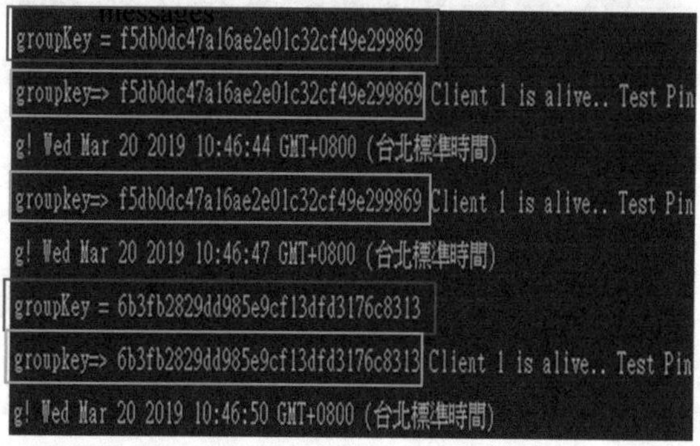

(d) The logged messages of a subscriber

Table 1 Hardware settings for Lab1 experiment in a LAN environment

	Client	Server
CPU	intel® Core™ i7-4790 CPU @ 3.60 GHz	intel® Core™ i7-4702MQ CPU @ 2.20 GHz
OS	Windows 7	Windows 10
RAM	6 GB	16 GB
Model	Acer Veriton M6630G	HP Probook 450 G1
Network card	Intel® Ethernet Connection I217-LM	Intel® Dual Band Wireless-AC 3160
Router	D-Link® DIR-809 Wireless AC750	D-Link® DIR-809 Wireless AC750
Software	Node.js 10.13.0, mqtt 2.18.3	Node 8.9.3, mongoose 5.4.1, mosca 2.8.3, passport-local 1.0.0

keys to respectively generate three encryptions for three subscribers. In the second experiment, the publisher encrypts the message, using the group key; the broker directly forwards the encryptions to the three clients, which decrypt the encryption, using the group key. We run each experiments more than 160 times, record the time between the publishing and the decryption at the subscribers. The average time of the first experiment takes 4.3 ms, and the average time of the second experiment takes 2.03 ms. We can see that the group-key-based solution only take 50% the time cost of the first experiment, even when there are only three clients. The improvement could be more significant when there are large number subscribers in the applications, as the broker needs to perform one individual encryption for each subscriber.

We summarize the merits of our MQTT group communication framework as follows.

- Device authentication with session key generation.
- Support secure group communications with automatic group key updating.
- Reduce the communication delay up to 50%, even when only three IoT devices are considered. The improvement would be much more significant when there are large number of devices.
- Significantly reduce computational overhead because brokers do not need to decrypt publishers' encryptions, and re-encrypt the messages again for subscribers.

5   Conclusions and Future Work

In this article, we have highlighted the importance of supporting group communications in the MQTT platforms. We have proposed our MQTT group communication framework and have implemented it as a prototype. We have conducted a simple three-subscriber-only experiments. The results show that the group-key-based solution takes only 50% the time of a conventional individual-encryption-based solution. As we can expect the improvements could be much significant when there are large numbers of subscribers in many practical applications. To evaluate the performance in a large field testing is one of our future works.

Acknowledgements This project is partially supported by the National Science Council, Taiwan, R.O.C., under grant no. MOST 107-2218-E-260-001 and Chunhua Su is supported by JSPS Kiban(B) 18H03240 and JSPS Kiban(C) 18K11298.

References

1. Avast: Avast research finds at least 32,000 smart homes and businesses at risk of leaking data. https://press.avast.com/avast-research-finds-at-least-32000-smart-homes-and-businesses-at-risk-of-leaking-data. Accessed 7 Nov 2018
2. MQTT: http://mqtt.org/. Accessed 7 Apr 2018
3. AMQP: Home. https://www.amqp.org/. Accessed 7 Nov 2018
4. CoAP—Constrained Application Protocol: Overview. http://coap.technology/. Accessed 7 Nov 2018
5. DDS Portal—Data Distribution Services—Object Management Group. https://www.omgwiki.org/dds/. Accessed 7 Nov 2018
6. ISO/IEC 20922:2016: Information technology—Message Queuing Telemetry Transport (MQTT) v3.1.1. https://www.iso.org/standard/69466.html. Accessed 7 Nov 2018
7. OASIS Message Queuing Telemetry Transport (MQTT) TC|OASIS. https://www.oasis-open.org/committees/mqtt/. Accessed 7 Nov 2018
8. Mirai (malware)—Wikipedia: https://en.wikipedia.org/wiki/Mirai_(malware). Accessed 7 Apr 2018
9. Amazon Web Services: Security and Identity for AWS IoT. https://docs.aws.amazon.com/iot/latest/developerguide/iot-security-identity.html. Accessed 17 Jan 2019
10. Mosquitto: http://projects.eclipse.org/projects/technology.mosquitto. Accessed 7 Nov 2018
11. Arduino cloud: https://cloud.arduino.cc/. Accessed 7 Nov 2018
12. Shiftr.io: https://shiftr.io/. Accessed 7 Nov 2018
13. Mosca: https://github.com/mcollina/mosca/. Accessed 7 Nov 2018
14. Chien, H.Y., Chen Y.J.: Security evaluation on various Arduino-compatible IoT devices. In: CISC2018, Taipei, 24, 25 May 2018
15. Andy, S., Rahardjo, B., Hanindhito, B.: Attack scenarios and security analysis of MQTT communication protocol in IoT system. In: Proceedings of EECSI 2017, Yogyakarta, Indonesia, 19–21 Sept 2017
16. Firdous, S.N., Baig, Z., Valli, C., Ibrahim, A.: Modelling and evaluation of malicious attacks against the IoT MQTT protocol. In: 2017 IEEE International Conference on Internet of Things (iThings) and IEEE Green Computing and Communications (GreenCom) and IEEE Cyber, Physical and Social Computing (CPSCom) and IEEE Smart Data (SmartData) (2017)
17. Shin, S.H., Kobara, K., Chuang, C.C., Huang, W.-C.: A security framework for MQTT. In: 2016 IEEE Conference on Communications and Network Security (CNS): International Workshop on Cyber-Physical Systems Security (CPS-Sec) (2016)
18. Shin, S.H., Kobara, K.: Efficient augmented password-only authentication and key exchange for IKEv2. IETF RFC 6628, Experimental, June 2012. https://tools.ietf.org/rfc/rfc6628.txt
19. Bhawiyuga, A., Data, M., Warda, A.: Architectural design of token based authentication of MQTT protocol in constrained IoT device. In: 2017 11th International Conference on Telecommunication Systems Services and Applications (TSSA), Lombok, Indonesia, 26–27 Oct 2017
20. Mektoubi, A., Lalaoui, H., Belhadaoui, H., Rifi, M., Zakari, A.: New approach for securing communication over MQTT protocol A comparison between RSA and Elliptic Curve. In: 2016 Third International Conference on Systems of Collaboration (SysCo), Casablanca, Morocco (2016)

21. Espinosa-Aranda, J.L., Vallez, N., Sanchez-Bueno, C., Aguado-Araujo, D., Bueno, G., Deniz, O.: Pulga, a tiny open-source MQTT broker for flexible and secure IoT deployments. In: 2015 IEEE Conference on Communications and Network Security (CNS), Florence, Italy, 28–30 Sept 2015

22. Rizzardi, A., Sicari, S., Miorandi, D., Coen-Porisini, A.O.: AUPS: an open source Authenticated publish/subscribe system for the internet of things. Inf. Syst. **62**, 29–41 (2016)

23. Lesjak, C., Hein, D., Hofmann, M., Maritsch, M., Aldrian, A., Priller, P., Ebner, T., Ruprechter, T., Pregartne, G.: Securing smart maintenance services: hardware-security and TLS for MQTT. In: IEEE 13th International Conference on Industrial Informatics (INDIN), Cambridge, UK, 22–24 July 2015

24. Arduino project: https://www.arduino.cc/. Accessed 7 Apr 2018

25. Raspberry pi: https://www.raspberrypi.org/. Accessed 7 Apr 2018

26. Arduino UNO wifi: https://www.arduino.cc/en/Guide/ArduinoUnoWiFi. Accessed 7 Apr 2018

27. Arduino MKR1000: https://www.arduino.cc/en/Main/ArduinoMKR1000?s_tact= C3970CMW. Accessed 7 Apr 2018

28. WeMos D1: https://wiki.wemos.cc/products:d1:d1_mini. Accessed 7 Apr 2018

29. Chien, H.Y., et al.: A MQTT-API-compatible IoT security-enhanced platform. submitted to the Int. J. Sens. Netw.

30. Introducing JSON: https://www.json.org/. Accessed 7 Nov 2018

31. NODE.JS: http://www.debugrun.com/a/cZomeQJ.html/. Accessed 7 Nov 2018

32. XMPP: About XMPP. https://xmpp.org/about/. Accessed 7 Nov 2018

33. Locke, D.: MQ Telemetry Transport (MQTT) V3.1 Protocol Specification. IBM Developer Works Technical Library, August 2010. http://www.ibm.com/developerworks/webservices/library/ws-mqtt/index.html

Author Index

© Springer Nature Switzerland AG 2020
R. Lee (ed.), *Computational Science/Intelligence and Applied Informatics*,
Studies in Computational Intelligence 848,
https://doi.org/10.1007/978-3-030-25225-0

Printed in the United States
By Bookmasters